Science and Sensibility

Gender and Scientific Enquiry,
1780–1945

Edited by
Marina Benjamin

Basil Blackwell

Copyright © Basil Blackwell 1991

First published 1991
First published in paperback in the UK 1993
First published in paperback in the USA 1994

Blackwell Publishers
108 Cowley Road, Oxford, OX4 1JF, UK

238 Main Street
Cambridge, MA 02142, USA

British Library Cataloguing in Publication Data

A CIP catalogue record for this book is available from the British Library.

Library of Congress Cataloging-in-Publication Data

A CIP catalogue record for this book is available from the Library of Congress.

ISBN 0631 19207 7

Typeset in 11 on 13 pt Caslon
by Hope Services (Abingdon) Ltd.
Printed in Great Britain by
T. J. Press Ltd., Padstow, Cornwall

Contents

Editor's Acknowledgements

I would like to thank the Twenty-Seven Foundation for their generous support of my research for this volume. Thanks are also due to Professor G. L'E. Turner for permission to quote from the Somerville papers, and to Virginia Murphy and Alison Dickens at Blackwell for their help at every stage of the book's preparation. This collection is the product of collaborative work, and I would like to extend thanks and appreciation to all the authors involved for their chapters, co-operation and enthusiasm: without them this book would not have been possible. I am particularly indebted to Ludmilla Jordanova, whose work is such an inspiration to students of gender history, for helping me gain the confidence to embark on this project, for her support and encouragement in its initial stages and her valuable comments on my contributions. I would also like to thank Roy Porter and Jim Secord for their generosity in lending their time and critical skills to readings of my work. Together, they gave me ever-higher standards to aim for. Numerous discussions with friends and scholars are all appreciated, and early conversations with Kate Cornwall-Jones, Andrew Cunningham, Anne Elder, Felicity Hunt, Steven Shapin and Perry Williams were especially helpful. Finally, thanks are due to my family, Bill Muir and Catherine Wearing for their moral support during this project's progress.

Contributors

MARINA BENJAMIN is writing a PhD on women, science and Romanticism 1780–1815. She is the author of several articles in the history of science, and is currently editing a forthcoming collection of essays on women, science and literature.

LYNDA BIRKE is a biologist, and is in the Department of Continuing Education at the University of Warwick. She is interested in the questions that science poses for women and for feminism, particularly in relation to biology. She is a founder member of the feminist group Women for Science for Women. Her recent publications include *Women, Feminism and Biology* (1986) and (with Susan Himmelweit and Gail Vines) *Tomorrow's Child: Reproductive Technology in the '90s* (1990).

ROGER COOTER is Senior Research Officer at the Wellcome Unit for the History of Medicine, University of Manchester, and one of the editors of *Social History of Medicine*. The author of *The Cultural Meaning of Popular Science* (1984); *Studies in the History of Alternative Medicine* (1988); and *Phrenology in the British Isles* (1989); he has also contributed many articles on the social history of science and medicine to books, journals and magazines. His *Surgery and Society in Peace and War: Orthopaedics and the Organization of Modern Medicine 1880–1948* is soon to be published.

GILL HUDSON is a PhD student at the University of Cambridge in the Department of History and Philosophy of Science, working on gender, science and pacifism in the 1920s and 30s. She is particularly interested in the interface between current theory in sociology of science and gender studies. Before taking up academic studies she followed a nursing career and brought up three children.

MARK MICALE is Assistant Professor of History at Yale University. From 1984 to 1987, he was a Junior Fellow at the Harvard Society of Fellows and from 1987 to 1989 a Post-Doctoral Fellow at the London Unit of the Wellcome Institute for the History of Medicine. His interests include modern European intellectual history and the history of medicine with an emphasis on the mental sciences in nineteenth-century France and Germany. He is currently working on a study of Charcot and the hysteria diagnosis in the nineteenth century, and is also preparing a translated and annotated edition of the historical essays of Henri Ellenberger.

A. D. MORRISON-LOW has been a curator of the history of science collections at the National Museums of Scotland since 1980. She has co-edited a collection of essays about the life and work of the nineteenth-century Scottish polymath, Sir David Brewster, and written a number of articles on the history of photography. Her interest in the structure of the scientific instrument trade outside London has resulted in contributions to two publications and she is co-author (with T. N. Clarke and A. D. C. Simpson) of *Brass & Glass: Scientific Instrument Making Workshops in Scotland as Illustrated by Instruments from the Frank Collection at the Royal Museum of Scotland* (1989) and (with J. Burnett) of *'Vulgar and Mechanick': The Scientific Instrument Trade in Ireland 1650–1921* (1990).

ORNELLA MOSCUCCI is the author of *The Science of Woman: Gynaecology and Gender in England, 1800–1929* (1990) and of a history of the Royal College of Obstetricians and Gynaecologists. She is currently working on a history of British obstetrics in the twentieth century.

LONDA SCHIEBINGER teaches European history at Pennsylvania State University and is the author of *The Mind Has No Sex? Women in the Origins of Modern Science* (1989). Her published articles include 'The history and philosophy of women in science: a review essay' in *Sex and Scientific Inquiry*, Jean O'Barr and Sandra Harding (eds), (1987); 'Skeletons in the closet: the first illustrations of the female skeleton in eighteenth-century anatomy' in *The Making of the Modern Body: Sexuality and Society in the Nineteenth Century* Catherine Gallagher and Thomas Laqueur (eds) (1987); 'Feminine icons: the face of early modern science' in *Critical Inquiry* (1988) and 'Maria Winkelmann and the Berlin Academy: a turning point for women in science' *Isis, Journal of the History of Science Society* (1987).

PERRY WILLIAMS is the Administrator at the Wellcome Unit for the History of Medicine, University of Cambridge. He has taught and written on various aspects of Victorian science and medicine, and he contributed the chapter on pioneer women university students in Felicity Hunt's *Lessons for Life*. He is now, with Andrew Cunningham, writing a popular account of the investigation of nature in the nineteenth century, entitled *The Invention of Science*.

Introduction

Marina Benjamin

In 1911, Charlotte Perkins Gilman, reflecting on the relation between the
sexes, wrote: 'To the man, the whole world was his world, his because he
was male; and the whole world of woman was the home because she was
female. She had her prescribed sphere, strictly limited to her feminine
occupations and interests; he had all the rest of life, and not only so, but
having it, insisted on calling it male.' Gilman, like many of her feminist
contemporaries, voiced a deep mistrust of the feminized world of
domesticity, which women had been instrumental in creating, but which
denied women's professional and political appetites, their capacity for
production as well as reproduction. Subsequent years have seen the
division of labour by gender – the notion of separate spheres of activity
for men and women in terms of the public and the private – become a
prime focus of interest for nineteenth-century social historians. Indeed,
we have learned that gender roles were perceived as God-given, respected
historically by tradition, and reinforced by social, political and legal
institutions. So fixed were these roles that Gilman could reflect with some
indignance and much justification, as Virginia Woolf and Simone de
Beauvoir were to do after her, that the condition she described was so
general and unbroken, that to speak of it 'arouses no more remark than
the statement of a natural law.'[1]

   Gilman's highlighting of the naturalization of gender roles is polemical
rather than analytical, amounting to a feminist indictment of the implied
fixity in men's and women's respective spheres of activity. It constitutes an
attack on traditionalists who sought to give, at the least a putatively
impartial, but often, a providential justification of their gendered social
hierarchy through an appeal to nature. But what of a critical analysis?

How did relations between the sexes come to be enshrined as natural laws? In other words, how were gender relations scientized? What relationships did women have with the sciences? And how can feminists interpret these relationships, actual and symbolic? These are the questions with which this volume is concerned.

THE LITERATURE ON WOMEN AND SCIENCE

The relevance of the natural sciences to major issues of concern within women's history has only recently been recognized. As a result, the marrying of women and science as a subject for scrutiny, has been transformed from the arcane to the topical, and is now sanctioned by a rapidly growing body of critical literature. Indeed, something of a new industry has been born of the recognition, on the one hand, that science is in the main patriarchal, and, on the other, that its interests are largely antithetical to feminist concerns. These realizations, have, for reasons which we will investigate, come late to the history of science; this has meant, however, that historians of science have been able to benefit from insights and methods now prevalent in broader areas of women's history. The identification of patriarchal interests, both in society and in the writing of its history, has provided feminist historians with justification for developing historiographical approaches which deliberately seek women's perspectives, and which thereby facilitate their attempts to reintegrate women's experiences into history. This revisionist history capitalizes on the understanding that the concept of gender, because it embodies that of difference, lends itself to social history as an analytical tool.[2] And gender analysis, by examining differences as they are manifested in the social world, has revealed how society nurtures inequalities between the sexes, and has demonstrated, reflectively, that gender, rather than being an immutable social category, is itself historically constituted.

The impetus behind employing gender analysis as a means of affording us a better understanding of our scientific heritage was provided by critical reappraisals of what biology meant to women. This project was fuelled in part by the rise of the Women's Health Movement, which challenged male medical control of women's bodies. The Boston Women's Health Care Collective, for instance, published a pioneering medical handbook for women, that has been updated several times since its appearance in 1971, and whose purposeful title, *Our Bodies Ourselves*, needs no elaboration.[3]

The debate over equal opportunites also made its mark, stimulating interest in equity studies that began to detail the ways in which the organization of scientific practice manifests gender inequalities; generally speaking, only a handful of women feature in head counts of scientific practitioners with prestigious or high-powered jobs, while the majority of women occupy low-status, unskilled and low or unpaid jobs.[4] Historians like Margaret Rossiter, questioning the received wisdom of what women can and cannot do, clearly demonstrate the imperative to address the visible roots of such gender stratification; the difficulties women face in gaining access to science education, training and practice, and the prejudices they encounter if they succeed. Rossiter's study of women scientists in America is a perceptive account of their 'struggles and strategies'.[5] Pursuit of this theme has revealed that the few women accepted into scientific professions have had to develop survival strategies to counter problems generated by their transgression of traditional female boundaries, and become hybrids — special cases in the laboratory and unconventional wives and mothers in the home.[6]

Historians have also focused on the more covert reasons behind the gendered stratification of scientific labour. Attention has been drawn to the prevalence of sexual metaphor within scientific discourses. Such metaphor serves symbolically to equate woman with nature as that which is known, as opposed to that which is capable of knowing. Carolyn Merchant has richly embellished this concept, showing the seventeenth century to have given birth in its mechanistic philosophy to a masculine science designed to unclothe, penetrate, dominate and conquer a female nature, variously portrayed as virgin, Venus, unruly whore and bountiful mother.[7]

The centrality of scientific discourses in the construction and representation of the female body and, by extension, female gender roles, has been fruitfully explored, especially in studies of the bio-medical sciences. We have recently learned from Ludmilla Jordanova that the power of science to objectify women is conveyed as much by images as by words. In teasing out the complex and subtle ways in which images of dissected women graphically embody the gendered relationship of knower to known, Jordanova reveals how the 'male gaze' of the anatomist unveils and penetrates the nature of woman; a nature represented as the very stuff she is made of, material and tangible, and opened out for viewing.[8] By considering issues of gender alongside those of ethnicity, we can discern how notions of female materiality were interlaced with scientific constructions of primitiveness. Sander Gilman's important study of racial stratification within nineteenth-century European theories of female sexuality centres on images of Hottentot women; science not only defined these women by their sexuality, characterizing them as more animal-like and lascivious than white women, but represented their bodies as pathologized versions of the feminine form.[9] For women of colour, access to the subject position of 'woman' was itself problematic.

Each of these studies adds force to the charge that the natural sciences, in their alliance with patriarchy, privilege men as a social category. Indeed, the charge of creating and enforcing the patriarchal social order has been extended beyond studies of women as subjects and objects of science, to analyses of the ways in which, through science, patriarchy received its epistemological consecration.[10] Interest in women and science has spread to women's history in general, so that even the roll-call of unsung heroines – fruit of the uncritical recovery of women – has been enriched by the inclusion of women scientists. Surely now only a few have not heard of Marie Curie, Caroline Herschel, or Sofia Kovalevskaia.

MAINSTREAM NEGLECT OF GENDER

Prior to this explosion in literature, relationships between women and science had been starved of attention by historians of science and feminist scholars alike. Rather than constituting a simple oversight, a blind dismissal, or a case of premeditated trivialization, this neglect is of a more fundamental nature. It is rooted in the fact that scientific practice has, since its origins, been almost exclusively male. It is worth noting in passing that Britain's oldest and most prestigious scientific institution, the Royal Society, did not open its doors to women until 1946, and that its French counterpart, the Académie des Sciences, only admitted women as members in 1979. Science is male in the dual sense that the natural sciences have been associated with men's work, and, moreover, with manly work.[11]

Women's history has shown us how such truisms should be challenged, but we must tread carefully, for the natural sciences are not homogeneous, and the degree to which different disciplines can be regarded as male varies both qualitatively and quantitatively. Indeed, botany was so much associated with female study that an article in *Science* for 1887 could reasonably ask, 'Is botany a suitable study for young men?' Its author, J. F. A. Adams, argued that botany should be studied by 'able-bodied and vigorous brained young men' as well as 'young ladies and effeminate youths'.[12] Yet it is the *maleness* of science that gender analysis is able to take to task and problematize, thereby creating room for discussing issues such as the feminization of specific disciplines. Hence it is not sufficient for our purposes to accept at face value that scientific practice has been and continues to be dominated by men, nor that its values coincide with what at any given time serves the interests of men. We must show how the identification of the natural sciences with men's work legitimates the exclusion of women from scientific practice, and that what is manly, far from being benign, is often in the service of exploiting women and marginalizing their interests. For gender analysis to make meaningful inroads into social history, it must confront the sense in which science is historically and materially male, that is, how, in real terms, it has accorded concrete benefits to men at the expense of women. It must examine the ways in which, by overlapping the interests of patriarchy and capitalism, science has operated to exclude women as practitioners and ideologically. And it must expose the extent to which science is acceptable because it replicates gender politics. We will see that gender, both as a language of sexual difference, and as a basis of power relations, is as central to the organization of scientific enterprise as it is to that of the workplace, domestic sphere, the worlds of politics and education – the concerns of the companion volumes in this series.

Here we meet with our first obstacle, because whilst conceptual problems pertaining to employment, the family, politics and education are situated firmly in the realm of social relations, those pertinent to science have customarily been articulated within a naturalist framework, that is, through an appeal to the natural world. Accounting for social processes such as exclusion, or the related activity of creating and sustaining hierarchical systems, has traditionally been low priority for historians of science. The adherence to a positivistic understanding of science led them to undermine the centrality of human relations in scientific theory and practice. Nineteenth-century positivists, like Comte and Mach, and members of the Vienna Circle, paraded scientific knowledge and the methods by which it is obtained as the ultimate achievement of human rationality. Science was represented, and indeed frequently presented itself, as the value-neutral activity of extracting and abstracting information from the natural world; its methods assumed a complete separation between subject and object, ostensibly ensuring that the knowledge they generated remained untainted by human influence. The scientist became a model of impartiality, a passive observer, collector and collator of natural facts; someone whose activities were unsullied by personal bias or ambition, unscarred by his political and religious standing, or by the cultural mores he upheld.[13] Social relations were banished from the laboratory. Moreover, the belief that science could directly access truth gained widespread currency in Victorian culture and meant that scientific methods could serve as a corrective to the subjectivity of the humanities. Thus science's sphere of influence was extended from explaining the workings of inanimate nature to unravelling the complexities of human nature.[14]

In recent years, the history of science has begun to divorce itself from its philosophical ancestry and to borrow its methodological underpinnings instead from sociology and anthropology.[15] Adopting from these disciplines a critique that rejects a clear-cut subject/object distinction and the value neutrality of natural facts in favour of an empirical relativism, the new historiography robs science of its sacrosanct status and renders it amenable to social analysis. The primacy of social relations to scientific activity is restored, and the scientist as social agent is removed from the ranks of the priesthood and transferred to that of the laity, where his desires for personal gain, glory, respect and influence all temper his activity inside as well as out of the laboratory. Sociologists and historians have focused on the contingent nature of scientific knowledge, that is, the processes whereby its character is shaped in the specific environments which generate it. They have spotlighted the ways in which the manufacture and utility of such knowledge serve the interests of particular social groups, whether these groups are characterized by their religious, political, economic, or other imperatives.[16] In short, the new sociology and history of science explode the positivist ideology of science, relegating its alleged purism to the realm of mythology; they reveal how scientific practice consists of precisely those social relations which it claims not to be contaminated by.

But while sociologists and historians embed the history of science securely within cultural history, they have proved as apt as the most traditional historians to neglect gender. The new literature curiously lacks a recognition of gender interests. With few exceptions, both the interests of women in science, and the interactions between science and patriarchy as symbiotic cultural constructs are ignored. Where the hierarchical structure of science has been emphasized, its patriarchal character has been obscured, much in the manner that gender has been subsumed in social history by class. It is blatantly not the case that the natural sciences have nothing to say about gender, or that women have been absent from their historical development. Quite the contrary; to take some examples, biology, psychology and psychoanalysis have been and continue to be preoccupied with gender, and women have long worked in science, albeit predominantly in low-status jobs, as technicians in physics and chemistry laboratories, astronomical data collectors, popularizers, disseminators and educators – not to mention jobs ancillary to science. Why then have those historians and sociologists of science who seek to problematize that which is taken for granted in science, failed to account for the most commonplace assumptions – those relating to gender difference?

A partial explanation is that the focus has been on elite, high-class science, the frontiers of discovery and controversy, areas where women are generally not to be found. Sociologists and historians have explored institutionalized science and neglected the domestic sphere. They have dwelt on the so-called 'hard sciences', like physics and chemistry, which indulge in less theorizing about women than the so-called 'soft sciences'.[17] Concentrating on high-class science not only masks women's interests in an enterprise where their activities are already handicapped by invisibility, it also, paradoxically, accords science the privileged status which the sociology of science seeks to challenge. And, focusing on the purportedly gender-blind hard sciences leads one to question to what extent the new historiography of the sciences itself shares the interests of patriarchy.

TOWARDS AN INTERDISCIPLINARY HISTORY OF SCIENCE AND GENDER

The current vogue for the study of women and science can thus be seen to be both a by-product and critique of contemporary histories of science. The recent interest in non-Western science and race-stratification in science shares these characteristics.[18] Although the mechanisms of race and gender are not mutually exclusive, the degree to which 'difference' within categories of gender is constituted and sustained by specific constructions of race, and vice versa, remains understudied. Historians of race and gender have instead concentrated primarily on separating the categories of race and gender away from the third component of the new theoretical trinity, namely, class.

To pursue this last point with reference to gender, we must recognize that while the concept of gender presupposes difference, and whilst, methodologically, the utility of gender analysis lies, as with class analysis, in its ability to interpret the manifestation and institutionalization of difference within hierarchical social orders, the conceptualization of gender as a function of class is a misleading one.[19] This is not to say that the two are unrelated, but to affirm that the relationship of male to female – the definition of these social categories with respect to one another and the continual policing of their separation – is negotiated and implemented in a manner distinct from the processes characterizing that of capitalist to worker. The notion that gender inequalities are secondary to or derivative from class inequalities induces us *a priori* into explaining women's activities and representation in terms of their oppression by men. With regard to science, such interpretations are not out of place, but neither are they sufficient.[20] We cannot afford to view women scientists solely as a category mistake, or judge science's objectification of women to be motivated primarily by the desire to repress them. If we did, then we would fail to register shifts in the capacity of particular scientific disciplines, like botany, to accommodate feminization, nor would we appreciate the means by which women managed to twist male-imposed ideologies of science to their own ends; women writing on science in the early nineteenth century, for instance, developed their own narrative language of science for women and children. The process of mapping the historical and cultural specificity of such instances of interfacing between science and gender enables us to transcend the dichotomy of oppressor/oppressed as the only basis for interpreting men's and women's respective relationships to science.

Although the study of women and science is steadily gaining mainstream acceptability, the literature remains strongly polarized between 'active' and 'passive' representations of women's relationship to science. Separate treatment is almost invariably accorded to women as subjects of science, that is, their active involvement as practitioners or audiences of science, and to women's inadvertent, passive or symbolic roles as objects of science. Finally, feminist concerns with science, and with women's experiences of science, have generally commanded a distinct place in the literature. While it is the case that these fields embody divergent historical traditions and conceptual problems, perpetuating their separation conceals the totality and complexity of the interactions between women, science and society. The existing literature on women and science fails to explore the interplay between the social and conceptual processes which are always working together to exclude women from science. It neglects, in the main, to stress how the scientific endorsement of distinct social roles for men and women justifies the structural barriers barring women from the ranks of practitioner, or how the persistent identification of women with the substrata of scientific investigation serves to alienate women from science, and prevent their desiring any fluency in the languages of science. The present volume comprises three sections — women practitioners of science, gender representation in science, and science and feminism — with the aim of combining instead of divorcing these areas. In this way, we can investigate how women's actual and conceptual involvements with the natural sciences illuminate each other, are mutually supportive, and have a common reliance on particular ideologies concerning the nature of woman, her capabilities and limitations; ideologies which take on full significance when placed in their historically specific contexts.

Each of the essays in this collection is a case study of a historically specific set of relations between women and science. It is no accident that the case study format has become the standard show-case for interdisciplinary history: focusing on a local context through a number of disciplinary lenses shows up assumptions which have long served as corner-stones of individual disciplines to be spurious. The case study checks, for example, the assumption commonly made by historians that a given term has a fixed meaning through time or a universal meaning across cultures. Case studies situated within both history of science and feminist studies reveal, among other things, that the term 'woman' is as contingent as the term 'science'. In short, local context shapes content. While, in this volume, the detailed context under scrutiny differs from one essay to another, there are some general aspects of nineteenth-century science which provide them with a common backdrop. The most important of these is the transformation of scientific practice from its Enlightenment role within polite culture as a gentlemanly pastime into a professional enterprise with a high degree of institutionalization.[21] Concomitantly, the sciences were secularized and harnessed into servicing the practical needs of industry rather than the metaphysical needs of theology. New disciplines such as physics, biology, geology, anthropology, gynaecology and psychoanalysis were named, and the widespread acceptance of positivism ensured that the scientific method they claimed to share was seen to be unrivalled as the route to true knowledge.

Though the twentieth century has witnessed the birth of 'big science' – the large-scale, high-powered and high-financed enterprise we know today, with its increasingly militarist orientation – it inherited much of the natural science's nineteenth-century form.[22] Indeed, the nineteenth-century processes which so effectively marginalized women's interests in science – the professionalization and institutionalization which rendered the laboratory a sanctum of masculinity, coupled to the increasing identification of woman as a domestic being, biologically unfitted for scientific pursuits – are still, some would argue, very much in evidence.

WOMEN PRACTITIONERS OF SCIENCE

The process of integrating women practitioners into the history of science has already undergone something of a revolution. Until very recently, research in this area was characterized by an impassioned but uncritical search for women who had been 'hidden from history'. Biographical narratives, unreflectively employed, put women on to the historical map as unsung heroines; for every great man of science, there emerged a great woman, whose merits had gone unrecognized because she had long been victimized, plagiarized, undervalued or simply ignored. Margaret Alic's *Hypatia's Heritage* (1986), has come, for its critics, to typify this genre; the unstated objective behind Alic's tendency to lionize scientific women, which in the context of feminist politics was already outdated, was to demonstrate that women too had achieved.[23] Though well intentioned, the attempt to mould women's activities to meet male criteria of success holds any genuine understanding of their careers in context at bay.[24] This applies equally when the women in question are understood to have failed to meet male standards. Dorothy Stein's recent biographical study of the life and mathematical work of Ada Lovelace, though thoroughly researched, lacks such awareness. Ada is set up as a scientific aspirant, only to have her ambitions read as pretentions.[25] Instead of trying to identify women's original contributions to science in terms of concepts or theories they succeeded or failed to 'mother', biographers now question traditional evaluations of what it means to be a 'successful scientist', thus shifting the focus away from the activities of individuals towards a more critical examination of gendered scientific hierarchies. From this perspective, issues of gender and authority, the division of labour by gender, and gendered ways of knowing become the parameters within which biographical accounts are situated.[26]

The chapters in this section deal with these and related issues, the most salient being the feminization of specific areas of science. Generally speaking, as scientific practice was increasingly organized along professional and industrial lines, so it was simultaneously gendered. Institutionalized science, growing up in the public sphere, evolved as a male domain. At the same time women's roles were increasingly defined in terms of the domestic sphere; even women in employment were encouraged to place priority on home and family life. Perry Williams and A. D. Morrison-Low, looking respectively at women and the sanitation movement and women in the scientific instrument trade, reveal, however, that the distinction between the public and the private sphere is easily blurred. By looking beyond institutionalized science and decentralizing our conceptions of what counts as scientific practice, Williams explores how the identification of sanitary education – a speciality lying without the domain of professional medicine – with philanthropy, enabled women to add the provision of such education to their repertoire of good causes. Similarly, we learn from Morrison-Low that because the scientific instrument trade, as a profession, retained a large craft-based element structurally, women's entry into the trade on the craft side in largely unskilled jobs was facilitated. She examines, moreover, the conditions which conspired to allow women some involvement in the management of scientific instrument companies; often women entered family businesses in a manner constructed as a natural extension of their role in the home, as their husbands' amenuenses, their sons' protectors, and as financial stewards. Marina Benjamin casts an eye back beyond the professionalization of science and explores how science could be gendered through its epistemological commitments rather than through formal divisions of labour; by relying on feminized cognitive processes, like feeling and sensibility, the very language of science carried implications for female sexuality. Benjamin looks at how women writers on science deployed scientific languages to articulate their views on female morality.

GENDER REPRESENTATION IN SCIENCE

The study of gender representation within science has produced a valuable literature of interdisciplinary character. Borrowing from literary criticism the use of literary analysis as method, it examines both the function of metaphor in science's personification of the natural world as female, and the importance of scientific languages in the construction of female gender roles. A central theme in this literature is the dialectical relationship between nature and culture. This pivots on the notion that man's development of natural philosophies, and subsequently natural sciences, suffused with sexual imagery, enabled him persistently to identify woman with nature and, in so doing, distance himself from nature through culture, which thereby became equivalent to science. Jordanova has explored the embeddedness of the nature/culture dualism in Enlightenment thinking, through a related set of dichotomies, like city/country and oppressor/oppressed; she shows that the identification of woman with nature, the country, materiality, passivity and sentiment was simultaneously a signifier of lack. Woman lacked rationality and agency, her spirituality tended towards superstition, she was everything that man-made culture sought to transcend.[27] The eighteenth century's insistent sexualization of the natural world, and the implications it held for social organization, is beautifully illustrated by Londa Schiebinger, who, in taking botany as her subject of investigation, tackles the concept of a masculine nature coexisting with female nature. Her chapter not only reveals that botanists customarily projected an anthropomorphic sexuality on to the plant kingdom, but, moreover, that the respectability of specific taxonomical systems depended largely on how reliably they reproduced the traditional gendered hierarchy found in society.

The investigation of gender representation in science is particularly pertinent to the nineteenth century, which witnessed the growth of what David Hume called the 'science of Man'. The seeds of secularization, sown initially in Enlightenment natural philosophy, are perhaps best exemplified in Scottish improvement philosophy, but natural philosophy soon followed suit. With the dawn of the nineteenth century, we can discern how scientific metaphor attained cultural or human rather than biblical significance, and how the languages of science, and for that matter literature, became vehicles for civic rather than Christian rhetoric. In other words, mankind became answerable to a secular science of Man. This development brought a confrontation with issues of gender difference into the limelight and gave rise to debates on the nature of woman, the most visible perhaps being those sparked off by social Darwinists. These debates serve to demonstrate how, in relation to gender, the social and symbolic dimensions of science interrelate symbiotically in that the scientization of woman's primary and secondary sexual characteristics directly fed domestic ideologies of womanhood. For example, Arthur Thomson and Patrick Geddes, who put forward their theory of the nature and evolution of sexual characteristics in *The Evolution of Sex* (1889), argued that on a metabolic level men were catabolic and therefore active, while women were anabolic and therefore passive.[28] Little extrapolation was then required to argue that women were biologically fitted for a passive role in society. Of course, the gender implications of scientific debates on femininity were culture-specific in that, for example, black women stood in very different relation to the ideals of womanliness.[29]

Three of the chapters in this section consider the medicalization of women against the backdrop of such debate. Focusing on the scientific discourses which constructed and deconstructed woman, body and mind, they explore the nineteenth century's preoccupation with uncovering the true nature of woman, of establishing scientifically precisely how she resembled yet differed from man. Roger Cooter reassesses the medical case history of Harriet Martineau. Going beyond the medical construction of woman, he explores, through Martineau's advocacy of mesmerism, the subversive ability of women to question the authority of the emergent male medical profession by throwing their support behind alternative therapies. Ornella Moscucci, situating her chapter within traditions of intellectual history, addresses the tensions arising from attempting at once to characterize both woman's equality to and difference from man. She examines conceptions of hermaphroditism, from a perspective that recognizes the potential of the science of Man and of evolutionary theory, in their stress on difference between the sexes, to undermine the unity of the species. This theme of comparative gender construction is taken up by Mark Micale, who analyses the relative construction of male and female hysteria. Micale details Jean-Martin Charcot's little-known work with male hysterics, and shows how this leading medical man's understanding of the causes and symptomatology of hysteria were coloured by prevailing social assumptions about gender roles. He suggests that, none the less, the British medical profession remained hostile to Charcot's theories; until the First World War they were reluctant to countenance any destabilization of masculinity, and, for them, hysteria remained an essentially female malady.

Science and Feminism

To subject the ideology of science to feminist criticism is to enter an arena of controversy and polemic, politics and ethics. For while there is consensus that science has contributed to men's domination of women, determining where and how the interests of science coincide with patriarchy is subject to as many permutations as there are schools of feminist thinking.

This field possesses a heterogeneous literature, but one with two principal focal points. First, the study of science's role in the licensing of gender stereotyping. The life sciences are a major target for criticism within such studies, mainly because they have had a high profile in prescribing specific social roles for woman, which are represented as being inevitable consequences of her biological and psychological make-up.[30] Such determinism operates in the interests of male domination, in that it legitimates only those activities of women which can be confined within biological parameters. Second, feminist criticism has exposed the mutually reinforcing relationship between science and patriarchy, broadly conceived, at its most fundamental level, that of epistemology. The central criticism here is that the need, so valued by scientists, for distinguishing between subject and object is essentially male. If, however, feminists argue that the masculine characteristics they identify are biological rather than cultural, the spectre of determinism is again raised. None the less, such studies have provided a launching pad for speculations concerning the epistemological underpinnings of what Sandra Harding terms 'successor sciences', and for visionary descriptions of how such would-be sciences might be practised.[31] Once again essentialism threatens, and feminists must be wary, in displacing patriarchal interests by their own, of universalizing their interests, and of thereby binding conceptualizations of femininity within alternative shackles. Lynda Birke, who looks at the issue of nature *versus* nurture in biology, confronts the problem of determinism head-on, arguing that we should resist temptations to treat feminist determinism as ideologically suspect. Regarded as a transitional formulation of contemporary concerns, it at least allows us to conceive of alternative modes of practising science. The main thrust of Birke's argument is that determinist theories are thoroughly compromised once we begin to understand human biology as transformation.

Gill Hudson, by contrast, deals with a very different feminist problem, namely, that of characterizing the relationship between women, science and pacifism. Her aim is twofold; to explore the links between masculinity and militarism, and, through a reading of the life and work of the pacifist scientist Dorothy Hodgkin, to show how her ideological commitments influenced the subject and mode of her research. By stressing women's experiences of science, this chapter provides an example of how discussions of woman-friendly science can be moved away from issues of epistemology and brought into the realm of social relations.

It should be noted that each of the chapters in this collection may be read in isolation. Together, however, they add to our still-limited understanding of both the scientific construction of gender and the gendered construction of science in the nineteenth century.

By way of conclusion, I want to return briefly to the idea of bringing different methodological approaches into confrontation with one another. The conceptual space created in the process is one which bears strong analogy to the ill-defined region midway between the poles of the nature/culture dichotomy – a region which Donna Haraway calls the 'borderlands'. Haraway invokes this term in her major new study of the history of primatology to describe the conceptual space inhabited by primates. The geographical metaphor is apt, since her work celebrates the opportunity that reading in the borderlands offers for 'story telling within several contested narrative fields', and of thereby overstepping the territorial boundaries established by any one discipline.[32] The value of reading in the borderlands lies in its highlighting the need to consult a wide diversity of historical sources from all areas of cultural activity. Such reading can be illuminated by considering the image on the jacket of the scientific-instrument trade catalogue (see plate 1). The advantage of using visual evidence to demonstrate a form of reading arises from the property of images to capture instantaneously a unity of expression from diverse meanings, which in fact operate on different levels. The image introducing this trade catalogue can be simultaneously situated within the history of scientific practice and within that of scientific representation, and, as such, it undermines a clear-cut distinction between the two.

This alluring image, rather than implying that science can analyse women, or that nature itself is female, serves a fundamentally different purpose, although it does affirm that science is manly. Like the use of women to adorn motor cars in today's advertising, the ethereal woman in this illustration, scantily dressed, and rising out of nothing like the genie

PLATE 1 Trade catalogue jacket for the High Tension Company Westminster.
(Reproduced by kind permission of the Trustees of the National Museums of Scotland)

of the lamp, was used to entice, to convey the message that science is sexy. This iconography of commercialism bears little relation, historically, to the female allegories of science popular in the seventeenth and eighteenth centuries.[33] Its purpose is to sell, and, moreover, to sell to men, since it blatantly assumes that scientific consumers are male. Indeed, the contents of this 1919 catalogue by the High Tension Company, Westminster, consist of a bald list of stock items (some illustrated) and prices. The desire both to sell instruments and at the same time enhance the masculine image of science is revealed in the deliberate ambiguity of the jacket illustration. The viewer cannot be certain that it depicts a scientific laboratory and not a boudoir; the heavy curtains in the background are luxuriously draped, and the sensual woman in the foreground is robed in evening wear, if not in a *négligé*. At the same time a large desk supports an induction-coil, and the floating female is holding a cathode-ray tube. The image equates man's quest for knowledge with the desirability of a woman, and a trade catalogue is the ideal place to symbolize this equation since it appeals directly to a desire for possession.[34] The link, now a commonplace, between electricity and sexuality is also graphically clear. It could, moreover, be argued that the almost Botticellian classicism of the female figure suggests that the instruments have a certain pedigree. Most interestingly, perhaps, the illustration distances itself from any pretence to realism; it is a painting rather than a photograph, featuring an idealized fantasy woman – perhaps even the spirit of nature – rather than a woman of flesh and blood, and it implies, crudely, that women cannot practise science.

The interface between the history of science and feminist studies can similarly be seen as a borderland, where meanings are made not given, and where the unquestioned is questioned. In this regard, feminist approaches have brought the history of science into line with the broad critique of modern Western culture which has preoccupied feminist historians for the past three decades. I think it needs to be spelt out, lest any doubts remain, that while borderlands represent margins they are not themselves marginal. Feminist historiographies of science, as one scholar has recently pointed out, do not simply supplement existing approaches and methods.[35] They carry the potential to unbalance the status quo, and to radically change the most deep-rooted beliefs about our scientific history.

NOTES

1 Charlotte Perkins Gilman, *The Man-Made World or our Androcentric Culture* (T. Fisher Unwin, London and Leipzig, 1911); Virginia Woolf, *A Room of*

One's Own (1929) (Grafton Books, London, 1989); Simone de Beauvoir, *The Second Sex* (1949) (Penguin, Harmondsworth, 1987).

2 The classic text on feminist historical method is Sheila Rowbotham's, *Hidden From History: 300 Years of Women's Oppression and the Fight Against It* (Pluto Press, London, 1973). For the current state of the art see Joan W. Scott, *Gender and the Politics of History* (Columbia University Press, New York, 1988).

3 Boston Women's Health Care Collective, *Our Bodies Ourselves* (Random House, New York, 1971). See also Claudia Dreifus (ed.), *Seizing Our Bodies: The Politics of Women's Health* (Vintage Books, New York, 1977); Barbara Ehrenreich and Deidre English *For Her Own Good: 150 years of Experts' Advice to Women* (Anchor Press, New York, 1978). The critical literature on biology includes: Ruth Hubbard and Marian Lowe (eds), *Genes and Gender II: Pitfalls in Research on Sex and Gender* (Gordian Press, New York, 1979); Ruth Hubbard, M. S. Henifin and B. Fried (eds), *Women Looking at Biology Looking at Women* (Shenkman Publishing Co., Cambridge, Mass., 1979); Brighton Women and Science Group (eds), *Alice Through the Microscope: The Power of Science over Women's Lives* (Virago, London, 1980); Donna Haraway, 'In the beginning was the Word: the Genesis of biological theory', *Signs*, 6 (1981), pp. 469–82; Janet Sayers, *Biological Politics: Feminist and Anti-Feminist Perspectives* (Tavistock Publications, and Methuen, London and New York, 1982); Ruth Bleier, *Science and Gender: A Critique of Biology and its Theories on Women* (Pergamon, New York and London, 1984); Ann Fausto-Sterling, *Myths of Gender: Biological Theories about Women and Men* (Basic Books, New York, 1985); Lynda Birke, *Women, Feminism and Biology, The Feminist Challenge*, (Wheatsheaf Books, Brighton, 1986).

4 See, for example: Johnathan Cole, *Fair Science: Women in the Scientific Community* (Free Press, New York; Collier Macmillan, London, 1979); Ann Briscoe and Sheila Pfafflin (eds), *Expanding the Role of Women in the Sciences*, Annals of the New York Academy of Sciences, 323 New York Academy of Sciences, (1979); Alison Kelly (ed.), *The Missing Half, Girls and Science Education* (Manchester University Press, Manchester, 1981); Sheila Humphreys (ed.), *Women and Minorities in Science: Strategies for Increasing Participation* (Westview, Boulder, 1982); Barbara Gross Davis, *Evaluating Intervention Programmes: Applications from Women's Programs in Maths and Science* (Teachers College, Columbia University, New York, 1985); Penina Migdal Glazer and Miriam Slater, *Unequal Colleagues: The Entrance of Women into the Professions 1890–1940* (Rutgers University Press, New Brunswick, 1987).

5 Margaret Rossiter, *Women Scientists in America: Struggles and Strategies to 1940* (Johns Hopkins University Press, Baltimore, 1982).

6 Sally Gregory Kohlstedt, 'In from the periphery: American women in science, 1830–1880', *Signs*, 4 (1978), pp. 91–6; Regina Markell Morantz-Sanchez, *Sympathy and Science: Women Physicians in American Medicine* (Oxford University Press, Oxford and New York, 1985); Pnina Abir-Am and Dorinda Outram (eds), *Uneasy Careers and Intimate Lives: Women in Science 1798–1979* (Rutgers University Press, New Brunswick, 1987); Londa Schiebinger, *The Mind has no Sex? Women in the Origins of Modern Science* (Harvard University Press, Cambridge, Mass., 1989).

7 Carolyn Merchant, *The Death of Nature: Women, Ecology and the Scientific Revolution* (Harper and Row, New York, 1980; London, 1982). See also, Brian Easlea, *Science and Sexual Oppression: Patriarchy's Confrontation with Woman and Nature* (Weidenfield and Nicolson, London, 1981); Evelyn Fox Keller, *Reflections on Science and Gender* (Yale University Press, 1985). For a critical reading of the Merchant thesis see Sylvana Tomaselli, 'The Enlightenment debate on women', *History Workshop Journal*, 20 (1985), pp. 101–24.

8 Ludmilla Jordanova, *Sexual Visions* (Harvester Wheatsheaf, Hemel Hempstead, 1989). See also Catherine Gallagher and Thomas Laqueur (eds), *The Making of the Modern Body: Sexuality and Society in the Nineteenth Century* University of California Press, California, 1987).

9 Sander L. Gilman, 'Black bodies, white bodies: toward an iconography of female sexuality in late nineteenth century art, medicine and literature', *Critical Inquiry*, 12, (1985), pp. 204–22. See also, Londa Schiebinger, 'The anatomy of difference: race and sex in eighteenth-century science', *Eighteenth Century Studies* 23, (1990) pp. 387–405.

10 A selection of the extensive literature on this subject includes: Sandra Harding and Merrill B. Hintikka (eds), *Discovering Reality: Feminist Perspectives on Epistemology, Metaphysics, Methodology, and Philosophy of Science* (Reidel, Dordrecht, Holland, 1983); Jan Harding (ed.), *Perspectives on Gender and Science* (Falmer Press, London, 1986); Ruth Bleier (ed.), *Feminist Approaches to Science* (Pergamon, New York, 1986); Sandra Harding, *The Science Question in Feminism* (Cornell University Press, Ithaca, and Open University Press, Milton Keynes, 1986); Sandra Harding and Jean F. O'Barr (eds), *Sex and Scientific Inquiry* (University of Chicago Press, Chicago, 1987).

11 An excellent example of how to analyse the embedding of masculinity in science and technology can be found in Cynthia Cockburn, *Machinery of Dominance, Women, Men and Technical Know-how* (Pluto Press, London, 1985). See also Fox Keller, *Reflections on Science and Gender*.

12 J. F. A. Adams, 'Is botany a suitable study for young men?' *Science*, 9 (1887), pp. 116–17. See Elizabeth B. Keeney, 'The Botanizers: Amateur Scientists in Nineteenth Century America' (Ph.D. thesis, University of Wisconsin, Madison, 1985). The institutional practice of botany, however,

remained closed to women – see D. E. Allen, *The Botanists: a History of the Botanical Society of the British Isles through a Hundred and Fifty Years* (St Paul's Bibliographies, Winchester, 1986).

13 See, for example, Auguste Comte, *A General View of Positivism* (1848), tr. Frederic Harrison (George Routledge and Sons, London, 1980); Robert S. Cohen and R. J. Seeger (eds), *Ernst Mach, Physicist and Philosopher* (Reidel, Dordrecht, 1970); Rudolf Carnap, *Logical Empiricist: Materials and Perspectives* (Reidel, Dordrecht, 1975); Ernest Nagel, *The Structure of Science: Problems in the Logic of Scientific Explanation* (1961) (Routledge and Kegan Paul, London, 1979); Maurice Mandelbaum, *Philosophy, History and the Sciences: Selected Critical Essays* (Johns Hopkins University Press, Baltimore and London, 1984); Otto Neurath (ed.), *Unified Science*, The Vienna Circle Monograph Series, tr. Hans Kaal, ed. Brian McGuiness, intro. Rainer Hegselmann (Reidel, Dordrecht and Lancaster, 1987).

14 See, for example, Frank Miller Turner, *Between Science and Religion: The Reaction to Scientific Naturalism in Late Victorian England* (Yale University Press, New Haven, 1974); Robert M. Young, *Darwin's Metaphor: Nature's Place in Victorian Culture* (Cambridge University Press, Cambridge, 1985); James R. Moore (ed.), *History, Humanity and Evolution* (Cambridge University Press, Cambridge, 1989); Theodore M. Porter, 'Natural Science and Social Theory' in R. C. Colby, G. N. Cantor, J. R. R. Christie and M. J. S. Hodge (eds), *Companion to the History of Modern Science* (Routledge, London and New York, 1990), pp. 1024–43.

15 Ludmilla Jordanova (ed.), *Languages of Nature, Critical Essays on Science and Literature* (Free Association Press, London, 1986): see introduction.

16 The methodology of the sociology of science is detailed in Bruno Latour and Steve Woolgar, *Laboratory Life: The Construction of Scientific Facts* (Sage Publications, Beverley Hills and London, 1979); Steven Shapin 'History of science and its sociological reconstructions', *History of Science*, 20 (1982), pp. 157–211; H.M. Collins and T. J. Pinch, *Frames of Meaning: The Social Construction of Extraordinary Science* (Routledge and Kegan Paul, London, 1982); Karin Knorr-Cetina and Michael Mulkay (eds), *Science Observed* (Sage Publications, London, 1983); Harry Collins, *Changing Order; Replication and Induction in Scientific Practice* (Sage Publications, London, 1985). For sociological case studies see the articles in *Social Studies of Science*, and in Barry Barnes and Steven Shapin (eds), *Natural Order: Historical Studies of Scientific Culture* (Sage Publications, Beverley Hills and London, 1979); T. J. Pinch, *Confronting Nature: The Sociology of Solar Neutrino Detection* (Reidel, Dordrecht and Lancaster, 1986); Bruno Latour, *The Pasturisation of France* (Harvard University Press, Cambridge, Mass., 1988); David Gooding, Trevor Pinch and Simon Schaffer, *The Uses of Experiment* (Cambridge University Press, Cambridge, 1989).

17 Sara Delamont, 'Three blind spots? A comment on the sociology of science by a puzzled outsider', *Social Studies of Science*, 17 (1987), pp. 163–70.

18 Leo Kuper (ed.), *Race, Science and Society*, (George Allen & Unwin, London, 1975) material first published by Unesco in 1956; Vijaya L. Melnick and Franklin D. Hamilton (eds), *Minorities in Science, The Challenge for Change in Biomedicine*, (Plenum Press, New York, 1977); Nancy Stepan, *The Idea of Race in Science; Great Britain 1800–1960* (Macmillan and St Antony's College, Oxford, London and Oxford, 1982); Willie Pearson Jr, *Black Scientists, White Society and Colorless Science, A Study of Universalism in American Science*, (Associated Faculty Press, Millwood, New York, 1985); N. Reingold and Marc Rothenberg (eds), *Scientific Colonialism, a Cross Cultural Comparison* (Washington, 1987); Lewis Pyenson, 'Science and imperialism', in Colby, Cantor, Christie and Hodge (eds), *Companion to the History of Modern Science*, pp. 920–33.

19 Judith L. Newton, Mary P. Ryan and Judith Walkowitz (eds), *Sex and Class in Women's History* (Routledge, London, 1983).

20 A recent reading of science principally as a tool in men's oppression of women is Cynthia Eagle Russet's, *Sexual Science: The Victorian Construction of Womanhood* (Harvard University Press, Cambridge, Mass., 1989).

21 See, for example: Jack Morrell and Arnold Thackray, *Gentlemen of Science: Early Years of the British Association for the Advancement of Science* (Clarendon Press, Oxford, 1981); S. F. Cannon, *Science in Culture: The Early Victorian Period* (Dawson, Folkestone, 1978); Adrian Desmond, *Archetypes and Ancestors; Palaeontology in Victorian London 1850–1875* (Blond and Briggs, London, 1982).

22 The standard work on 'big science' remains Derek J. de Solla Price's, *Little Science, Big Science* (Columbia University Press, New York and London, 1963). Some recent examples include: Andrew Pickering, *Constructing Quarks* (Edinburgh University Press, Edinburgh, 1983); Peter Galison, *How Experiments End* (Chicago University Press, Chicago, 1987); Armin Hermann, John Krige, Ulrike Mersits and Dominique Pestre, *History of CERN. Volume I. Launching the European Organisation for Nuclear Research* (North Holland, Amsterdam, 1987); David A. Hounshell and John Kenly Smith Jr, *Science and Corporate Strategy: Du Pont R & D, 1902–1980* (Cambridge University Press, Cambridge, 1989).

23 Margaret Alic, *Hypatia's Heritage, A History of Women in Science from Antiquity to the Late Nineteenth Century* (Women's Press, London, 1986); see also Dorinda Outram's review of Alic in *The British Journal for the History of Science*, 20 (1987), pp. 224–5.

24 Sylvana Tomaselli, 'Collecting women: the female in scientific biography', *Science as Culture*, 4 (1988), pp. 95–106.

25 Dorothy Stein, *Ada: A Life and a Legacy* (MIT Press, Cambridge, Mass., 1985).

26 Some fine recent scientific biographies include: Louis Bucciarelli and Nancy Dworsky, *Sophie Germain: An Essay in the History of the Theory of Elasticity* (Reidel, Dordrecht and London, 1980); Elizabeth Chambers Patterson, *Mary Somerville and the Cultivation of Science 1815–1840* (Martinus Nijhoff Publishers, The Hague, 1983); Ann Hibner Koblitz, *A Convergence of Lives, Sofia Kovalevskaia: Scientist, Writer, Revolutionary* (Birkhauser, Boston, 1983); Evelyn Fox Keller, *A Feeling for the Organism: the Life and Work of Barbara McClintock* (W. H. Freeman, New York, 1982); Francoise Giroud, *Marie Curie, a Life*, tr. Lydia Davis (Holmes and Meier, New York, 1986); see also the essays in part II of Abir-Am and Outram, *Uneasy Careers and Intimate Lives*; Kathleen Jones, *A Glorious Fame, the Life of Margaret Cavendish, Duchess of Newcastle, 1623–1673* (Bloomsbury, London, 1988). For a history of biographies on women scientists, see Londa Schiebinger, 'The history and philosophy of women in science: a review essay', *Signs*, 12 (1987), pp. 305–32.

27 By far the best exposition of this approach, and example of its use, is Ludmilla Jordanova's 'Natural facts: a historical perspective on science and sexuality' in Carol MacCormack and Marilyn Strathern (eds), *Nature, Culture and Gender* (Cambridge University Press, Cambridge and New York, 1980). See also the essays in Ruth Hubbard and Marian Lowe (eds), *Woman's Nature: Rationalisations of Inequality* (Pergamon, New York, 1983); cf. with references in n. 6 above.

28 See Elizabeth Fee, *Science and the 'Woman Question' 1860–1920* (Ph.D. thesis, Princeton University 1978, University Microfilms International, Ann Arbor, Michigan and London, 1978).

29 For discussion of gender-identity formation in the context of American slavery see Hazel V. Carby, *Reconstructing Womanhood, The Emergence of the Afro-American Woman Novelist*, (Oxford University Press, Oxford, 1987) and Hortense J. Spillers, 'Mama's baby, papa's maybe: an American grammar book', *Diacritics*, 17, no. 2, (1987), pp. 65–81.

30 Cf. with nn. 3 and 6 above.

31 Harding, *The Science Question in Feminism*. For discussions of feminist epistemologies for the sciences see, among others, Evelyn Fox Keller, 'Feminism and Science', *Signs*, 7 (1982), pp. 589–602; Madeleine J. Goodman and Lenn Evan Goodman, 'Is there a feminist biology?' *International Journal of Women's Studies*, 4 (1981), pp. 393–413; Hilary Rose, 'Hand, brain and heart: a feminist epistemology for the natural sciences', *Signs*, 9 (1983), pp. 73–90. Nancy Goddard and Mary Sue Henifin, 'A feminist approach to the biology of women', *Women's Studies Quarterly*, 12 (1984), pp. 11–18; Sue V. Rosser, 'A call for feminist

science'; *International Journal of Women's Studies*, 7 (1984), pp. 3–9; Janet Sayers, 'Feminism and science – reason and passion', *Women's Studies International Forum*, 10 (1987), pp. 171–9; Margareta Halberg, 'Feminist epistemology: an impossible project?' *Radical Philosophy*, 53 (1989), pp. 3–7.

32 Donna Haraway, *Primate Visions, Gender, Race and Nature in the World of Modern Science* (Routledge, New York and London, 1989); quotation from p. 6.

33 Londa Schiebinger, 'Feminine icons: the face of early modern science', *Critical Inquiry*, 14 (1988), pp. 661–91.

34 For discussion of association between scientific knowledge, desire and possession see Jordanova, *Sexual Visions*.

35 John Christie, 'Feminism in the History of Science' in Colby, Cantor, Christie and Hodge (eds), *Companion to the History of Modern Science*, pp. 100–9.

PART I

Women Practitioners of Science

1

Elbow Room:
Women Writers on Science, 1790–1840

Marina Benjamin

Olivia Smith's recent study of the social and political divisiveness of language in the period 1791–1819 is a fascinating elaboration of the association of refined language with breeding, civilization and morality, and of the vernacular with licentiousness, rusticity and barbarity.[1] Though dealing principally with the formal structure of language, it serves as a useful springboard for a discussion of the gender divisiveness of British science as public discourse, at a time when the public sphere was diligently policed by a State eager to stem any flow of republican sympathy towards revolutionary France. The model of refined versus vernacular language, and the implication it carries that language conveys cultural meanings beyond its explicit content, can be transposed to accommodate an illustration of the tensions between different modes of scientific discourse. Such tensions will be examined principally in relation to gender, not least because the cognates of the terms 'refined' and 'vernacular' were themselves gendered, though not in any simple fashion. While the extended meaning of the term 'vernacular' can be seen to have included 'woman', it did not include all things womanly; for example, sensibility – a distinctly feminized quality – denoted both refined and vernacular, both woman's glory and downfall.

   In aiming to uncover the gender ideology encoded in scientific languages which ostensibly have nothing to say about women, I will attempt to draw out the implications for female morality of prescribing specific ways of knowing nature, mainly concentrating on the significance of single terms.[2] Feminist thinkers have already drawn attention to the masculine character of scientific epistemologies, concentrating on the models of knowing which implicate dichotomies like rational/emotional, deductive/intuitive and objective/subjective, and align them with the dichotomy masculine/feminine.[3] This chapter will explore the ways in which such an approach to scientific discourse can be usefully extended and historicized, since the gendered nature of such discourse is not always unambiguous or oppositional, and concepts which no longer typify the way we think about science have in the past had purchase on both science and sexuality. The languages of science in the late eighteenth and early nineteenth centuries are a case in point. Here, disquisitions on the cognitive role of the passions, imagination, sensation and individual experience constituted the terminological hinges of debate; these were subjects central both to conflicts about the proper means of acquiring knowledge and to mooting the morality of sexual expression.

To focus on women writers on science, that is, women engaged in public discourse, is to look at women as protagonists of cultural activity, as agents not objects of civilization, and thus of their own fate. Such a perspective is not as obvious as it may appear, and I am indebted to those scholars who have indicated the extent to which privileging gender brings into question the assumptions usually brought to the writing of history.[4] In our particular concern with the sciences, we should note at the outset that the period under consideration saw little challenge by women to the intellectual authority or social status of scientific men through a direct involvement with science. Yet women's limited public dialogue with men of science does not exonerate those historians of science, who, mistaking the predominance of the male voice in scientific discourses for the sole voice, have effectively masked women's involvement with scientific practice altogether. We need not look far to find justification for looking at women writers on science. Overtly, and somewhat crudely, in terms of numbers, more women engaged with science in thought than in deed, through commentary not innovation. Their writing activity, seen as a contribution to a burgeoning corpus of women's educational treatises, can be set in the context of one of the few accepted pairings of woman and culture, namely pedagogy. On a more covert level, if we accept that they were addressing female manners and morals, these women, by entering the discourse of male science, were highlighting the affiliations between science and sexuality through their use of a shared vocabulary. This is a notion which deserves greater attention, since by definition to share is to cut both ways, and we will see that as women were embracing science within a language of female manners and morals, men were addressing sexuality through their science.

ROMANTIC SCIENCE AND FEMALE SEXUALITY

'For what purpose were the passions implanted?' Mary Wollstonecraft asked rhetorically in 1792, and replied, 'That man by struggling with them might attain a degree of knowledge denied to the brutes.'[5] Romantic natural philosophers would have answered differently. To speak meaningfully of Romantic science, some care must be taken in conveying what is meant by the term 'Romanticism', for as has been frequently acknowledged, there was no single Romanticism, no unified pan-European movement, or school of thought, deserving sole claim to the appellation 'Romantic'.[6] However, there are some general themes with which the term is most often associated which will concern us; the central preoccupation with the creative role of the imagination, the stress on the validity of an individual and an individual's sensory perceptions, the valuing of the natural over the artificial, and the search through self-knowledge for universal truths about the social or natural world.[7] I will use the phrase 'Romantic science' in a general sense to refer to modes of natural inquiry which shared these perspectives. Romantic natural philosophers eschewed the Enlightenment rationalism – institutionalized in the universities and the Royal Society – which celebrated diligent observation and notebook recording of external phenomena as the foundation of natural knowledge. They resurrected Bacon's idols of the cave and made the subjective experience of the philosopher speak for realism.[8] *Contra* Wollstonecraft, these scientific practitioners happily obeyed rather than battled with the dictates of their passions, privileging information originating in the mind and not in nature.

This faith in the mind's impressions drew support from a vitalist physiology, which analysed human sensation as the end product of organized matter. Theories of sensibility, or of genius, served to legitimate natural philosophies that scrutinized the active powers in mind and matter.[9] It followed that the Romantic natural philosopher adopted a wide range of poses extending from willing victim of natural power to, at the other extreme, manipulative master of nature's active agents. Both can be seen as variations on a theme derived from Rousseau, that is, the notion that man could gain access to an uncorrupt primitive self by abandoning the artifice of society for an internalized natural world of sensations. And in many cases these poses were literal. In the 1770s John Walsh described in detail in a series of letters addressed to Benjamin Franklin the shock, then torpor, arising from his experiences of electrocution by the torpedo fish. His experiments prompted Cavendish to manufacture an artificial torpedo which dispensed electric shocks, and fed Joseph Priestley's speculations concerning the identification of nervous fluids with common electricity.[10] Pneumatics as well as electricity presented fertile ground for auto-experimentation. At the Pneumatic Institution at Clifton, Bristol, in 1799, the chemist Humphry Davy, and Thomas Beddoes, the clinic's founder, along with the poets Coleridge and Southey, were transported by the intoxicating effects of nitrous oxide. Inhalation became the poetic analogue of inspiration.[11] Davy enthused, 'I existed in a world of newly connected and newly modified ideas, I theorised, I imagined that I made discoveries. . . . Nothing exists but thoughts! The universe is composed of impressions, ideas, pleasures and pains!'[12]

Interest in natural agents can be linked to a materialist tradition in natural philosophy which dates from the mid-eighteenth century, and which became pervasive through the widespread adoption of the ideas of thinkers such as David Hartley. As David Knight has suggested, this interest in powers and forces may usefully be called 'pre-Romantic'.[13] In Davy's case, this tradition was worked into a new framework, for no less significant in his world view were the aspirations he shared with the Lake poets whom he met at Bristol. The youthful radical cosmology of Wordsworth and Coleridge, expressed in their *Lyrical Ballads* (1798), which promoted natural simplicity and vernacular language at the expense of civilized values and artifice, found an echo in Davy's own nature poems, which at this time verged on the pantheistic. Davy's inaugural public lecture at the Royal Institution in 1802, published in pamphlet form in the same year as *A Discourse, Introductory to a Course of Lectures on Chemistry*, further took up the themes of the *Lyrical Ballads*. This showed that chemistry, like poetry, made the natural world legible through its stress on unifying principles and affinities, and its rejection of elegant hypotheses. Nor was the imagination, which Wordsworth aligned with pure reason, dispensable; Davy described the attempt to define the general laws governing phenomena as being 'lost in obscure, though sublime imaginations concerning unknown agencies'.[14] Thus the creative natural philosopher could succeed in making nature readable, and thereby accessible, as the popularity of Davy's lectures in London attested.[15]

Demonstrating control of natural power through auto-experimentation gave rise to a related set of practices. The public were treated to instructive electrical and pneumatic performances by itinerant lecturers and to extravagant displays of the philosopher's power by magnetizers, galvanizers and physiognomists. These spectacles at once served to evidence divine intervention in the natural world, and demonstrate the conviction, so well articulated by Davy in his *Discourse* of 1802, that science had bestowed upon man 'powers which may be almost called creative; which have enabled him to modify and change the beings surrounding him, and by his experiments to interrogate nature with power, not simply as a scholar, passive and seeking only to understand her operations, but rather as a master active with his own instruments.'[16] The links between power, natural knowledge and masculinity underlying the above passage will be considered later with reference to scientific representation in Mary Shelley's *Frankenstein*.

Mastering natural power was no less fundamental to therapy than to instruction. Popularizers of animal magnetism presented themselves as mystical communicators of a vital force, which, once confined to the mesmeric wand or 'tub', could revitalize the human spirit, serving most notably to counter the *ennui* of the *ancien régime*. Mesmer conceived animal magnetism as a universal, imponderable fluid, a material life-force akin to the Newtonian aether.[17] Electrotherapists too sought to anchor the beneficial effects of their treatments in theories that equated nervous fluid with a vital electrical force seen to be analogous to atmospheric electricity.[18] Significantly, for our concerns, most of the patients were women. Aside from conveying improvements in general health and relieving the trivial symptoms of fatigue and *ennui*, these unconventional therapies had particular application to female ailments, menstrual problems, infertility and hysteria. This claim was more often implied than stated, and is most conspicuously alluded to by the illustrations to the large array of texts on these treatments, which almost invariably represented the patients as women. Indeed, Mesmer himself, describing his cure for one of his first patients, Fräulein Oesterline, who manifested a 'convulsive malady', concluded somewhat euphemistically by affirming, 'She married and had children.'[19] It appears, then, that man's control of women's reproductive capacity was an integral aspect of the natural philosopher's mastery of nature.

The speculative subjective sciences of discovery had always sat uncomfortably beside the modes of natural inquiry brought together under the umbrella of the academies, notably the Royal Society and the Académie des Sciences. This tension surfaced in Davy's writings when he argued that in the search for discovery, 'The man of true genius . . . will rather pursue the plans of his own mind than be limited by the artificial divisions of language.'[20] Moreover, the popular, not to mention financial, success of mesmerism, galvanism, physiognomy, electrotherapy and pneumatics threatened the establishment's monopoly on medical practice and invited response. In 1784 the Faculty of Medicine in Paris organized by royal appointment a commission to investigate the activities of Mesmer and his followers, many of whom were already alienated from the academies. Jean Sylvain Bailly's report to the Faculty denounced the traders as charlatans and corruptors of morality, and animal magnetism as chimerical. According to the commission, the therapeutic effects of mesmerism derived solely from the imagination – a conclusion affirmed in an independent investigation conducted by the Académie des Sciences. Damning the faculty of imagination was a powerful resource for critics of Romantic science on an epistemological level, since it exploited the difficulty of corroborating and validating an individual's experience. In the revolutionary·climate of the 1790s such accusations took on a more distinct political significance, and it suited reactionary critics to lump together mesmerism, electrotherapy and galvanism as a single target for attack. These philosophies of nature were seen to be implicated in subversive plots to undermine the existing social order; Mesmer's secretive and exclusive Society of Harmony was aligned with Adam Weishaupt's radical sect the 'Illuminati', and all were linked to other heterodox brands of spiritualism seen to be distanced but a hair's breadth from atheism: Swedenborgianism, Rosicrucianism, pantheism and millenarianism.[21] Even the commonly devout practice of physiognomy, the reading of an individual's nature through his or her physical characteristics, was suspect – a likely reflection of the success of reactionaries like Edmund Burke in equating the popular with the populist, and novelty with innovation.[22]

For women, who were thought to be more passionate and to have a greater sensibility than men, the Romantic sciences were no less an evil than Pandora's box, promising moral and sexual (these categories were not always distinguishable) danger and temptation. In an attack on animal magnetism and the quackery of Mesmer's British disciples, principally De Mainauduc, the Baptist minister, John Martin pondered the crimes to which unsuspecting mesmerized ladies who 'are not responsible for their behaviour' were exposed.[23] Somnambulism, like the drug-induced transports of Davy, or the opium eating of Coleridge and De Quincey, was a symbol of the Romantic abandon of self to sensory experience. But for women this form of liberation and the suggestibility it implied was less a matter of exploring unusual states of consciousness than a foolhardy act of immorality. Later it was possible for women to subvert the politico-moral meanings attached to somnambulism, as Harriet Matineau's advocacy of mesmerism, which Roger Cooter discusses below, illustrates. The fundamental threat that early Romantic sensualism posed to sexual propriety lay in its stress upon the imagination. Hannah More, the evangelical polemicist, who condemned mesmerism and physiognomy as 'demoniacal mummery',[24] warned against succumbing to this powerful faculty when she censured those arts which called it into play: 'The imagination thus excited, and no longer under the government of strict principle, becomes the most dangerous stimulant of the passions.'[25] From the mid-eighteenth century, the imagination was held responsible not only for sexual passion, but for a whole range of sexual maladies, from vulgar eroticism to nymphomania.[26] To gain some idea of women's sensitivity in the censorious climate of the 1790s to the liberating properties of creative scientific imagination we will briefly consider a few examples from literature and from pedagogy – areas where women had some public presence.

The Gothic novels, immensely popular in this decade, possess features that are germane to our concern with science and gender. A number of commentators have convincingly argued that the Gothic novel is an allegory of revolution, that a critique of arbitrary power is contained in narratives of an innocent heroine's struggle for self-identity against domination by a corrupt and powerful man.[27] In other words, this Gothic format is profoundly radical. The republican Charlotte Smith co-opted the discourse of imaginative science into the political sub-text of her novels. The independence-seeking heroines of *Desmond* (1792) and *Marchmont* (1796) are both talented physiognomists. It is interesting to note that the radical Thomas Holcroft, author of *Anna St Ives*, was translator of a popular edition of Lavater's *Essays on Physiognomy*.[28] In a more general sense, thunder and lightning, electric shocks, occult forces, hallucinations and ghosts were the foodstuff of Gothic novels. England's foremost Gothic novelist, Ann Radcliffe, invoked these trappings in *The Romance of the Forest* (1791) and *The Mysteries of Udolpho* (1794) as the backdrop against which her heroines struggled for self-identity. Her technique was to use the heroine's realization that these seemingly supernatural effects had a natural basis to symbolize her liberation from her persecutors. One scholar has suggested that this process of realization was simultaneously an exploration of the secret regions of female desire, so that the heroine's independence from male domination involves her coming to terms with her sexuality.[29] Radcliffe, whose plots characteristically centred around the consciousness and sensations of her young impressionable heroines, fell silent after publishing *The Italian* in 1797. If we are persuaded that the genre helped gel an association between female sexual liberation and radical politics, then we have a plausible reason for interpreting why the politically conservative Radcliffe censored herself during the height of counter-revolutionary fever.

The association between sexual and political liberation was frequently invoked in a Britain preoccupied with the need to regain the political stability which had been so severely undermined by the French Revolution. In imitation of Burke, who represented the momentous event over the Channel as a rape, sexual violation became a common metaphor for revolution. Coleridge, for instance, in his poem *Happiness* (1791) linked revolution to unbridled desire.[30] These associations should not simply be seen as metaphoric, since the negotiation of gender roles is an inherently political process, linked intimately to debates concerning individual identity and the role of the individual in society. And the impact of the revolution in France placed the process of defining liberty and its implications for the structuring of civilized society on the centre stage of British culture. Mary Wollstonecraft, celebrated for *A Vindication of the Rights of Woman* (1792) – one of her contributions on the republican side to this debate on liberty – further forged the link between female sexuality and radical politics, albeit somewhat inadvertently. R. M. Janes has shown that the *Vindication* was initially well received by the critics, the tide of public taste turning only after William Godwin published his *Memoirs of the Author of A Vindication of the Rights of Woman*, which elaborated a history of Mary's unconventional love-life.[31] In the public eye, subversive politics was seen to have entered the boudoir in the guise of sexual licentiousness.

In the same year that Godwin brought out his *Memoirs*, the Reverend Richard Polwhele, a prominent Anti-Jacobin, published an anti-Wollstonecraftian poem entitled *The Unsex'd females*. Here, Mrs Barbauld, Charlotte Smith, Mary Hays, Helen Williams and Angelica Kauffman are all lined up for ridicule, accused of exhibiting 'Gallic mania'. This poem is a good example of how criticism of sexual and political liberalism could be articulated through an attack on Romantic science. The style and content of the poem parody Erasmus Darwin's epic poem *The Botanic*

Garden (1789–91), where the libertine Darwin indulged in a visionary and eulogistic tribute to the sexual behaviour of the vegetable kingdom. Whether or not Darwin's scientific work can be related to the ideas of the early Romantics has been discussed elsewhere, here our concern is with Polwhele, who made the most of such a relationship.[32] Using Bailly's disclaimer, Polwhele jibed that while Darwin may have been aware that the imagination refused to enlist under the banner of science, 'yet science may sometimes be brought forward not unhappily under the conduct of imagination'; and with *The Botanic Garden*, 'we are presented with a complete specimen.' Polwhele, reading Darwin as legitimating a free-loving licentiousness and disregard of moral probity in the social world, claimed he could not reconcile women studying 'the sexual system of plants' with 'female modesty'. Even Emma Crewe's frontispiece to *The Loves of Plants* had 'an air of voluptuousness too luxuriously melting'. He made his point quite clear by titillating the reader with a lustful language of sexual innuendo, and remarked, 'To such language our botanising girls are doubtless familiarised: and, they are in a fair way of becoming worthy disciples of Miss W.'[33] As Londa Schiebinger observes below, Polwhele, like Darwin, employed languages of science to make political points about female sexuality.

The subject of female decorum was not far from the surface of the condemnation of Romantic science which the didactic writer Maria Edgeworth effected through a comparison with chemistry:

> Chemistry is a science particularly suited to women, suited to their talents and their situation. Chemistry is not a science of parade, it affords occupation and infinite variety; it demands no bodily strength, it can be pursued in retirement, it applies immediately to useful and domestic purposes; and whilst the ingenuity of the most inventive mind may be exercised, there is no danger of inflaming the imagination; the judgement is improved, the mind is intent upon realities, the knowledge that is acquired is exact, and the pleasure of the pursuit is a sufficient reward for the labour.[34]

Though Edgeworth shunned Romantic science, her political sympathies were decidedly liberal. The same is true of Mary Wollstonecraft, for whom mesmerism was nothing but 'hocus pocus tricks', performed by 'priests of quackery' not for the love of God but money.[35] There is much in the above passage from *Letters for Literary Ladies* (1795) that therefore requires comment. Like her eccentric inventor father, Richard Lovell Edgeworth, Maria moved in dissenting and scientific circles which fostered progressive views towards women's education and social reform. The heroines of her novels were not the fanciful innocent creatures who populated the pages of the day's sentimental novels, but independent, responsible and useful women, who, like herself, placed special priority on issues of pedagogy. Utility was her guiding light, and this was precisely what she deemed lacking in the existing orientation of female studies. *Practical Education* (1798), which Edgeworth co-authored with her father, served as a platform for a critique of aristocratic manners and morals which she maintained had their source in a defective schooling in ornamental accomplishments. Instead of needlework, modern languages, dancing and drawing, she championed chemistry, botany and mechanics as useful study for women. These sciences depended upon the 'ingenuity' not sentiment of the student and led to an 'exact' knowledge of 'realities'. Edgeworth's scientism was a persistent theme throughout her writings, and it is to this mode of scientism which we will now turn.[36]

CONSERVATIVE SYSTEMATICS AND THE SOCIAL ORDER

Opposed to the subjective sciences that drew on Romantic culture were the descriptive sciences of natural history and the mathematical sciences like astronomy, where the academic replaced the dilettante, scholarship replaced discovery, and judgement replaced genius or sentiment. Natural philosophers patiently observed, detailed and classified the natural world looking for beauty, order, harmony and hierarchy, and referring causes to God's design, wisdom and benevolence. Conservative thinkers grounded these philosophies of nature on the argument from design — centre-point of an age-old natural theology newly dressed in the turbulent 1790s by William Paley. Natural theology was the corner-stone of Paley's moral philosophy, providing primary evidence of benevolent contrivance in a cosmology predicated on the belief that the security of civil society depended upon an appreciation of its providential basis.[37] In contrast to subversively manipulating active powers, the task of the philosopher was to discern divinely imposed systems in nature, so that natural knowledge became godly knowledge. The religious imperative was crucial; 'Direct experience', as Simon Schaffer has indicated, 'would lead to *superstition*, as it had for primitive peoples; systematic and progressive experience led to *rational beliefs*.'[38] The ethical agenda too was appropriately conservative, for systematics demanded that the imagination be made redundant, morality depended upon the refinement of understanding, not upon vulgar feeling.

Systems served as a resource for conservative natural philosophers in their campaign to discredit alternative sciences on political grounds. In a 1797 pamphlet, John Robison, Professor of Natural Philosophy at Edinburgh, branded the republican politics, irreligion and materialism of Priestley and his followers as the related products of illuminism.[39] While Robison alerted the public to subversive modes of scientific practice, the natural historian Sir Joseph Banks, as President of the Royal Society, found practical ways of barring radicals from entry into the world of prestigious science; he effectively withheld his patronage from radicals like Thomas Beddoes and Thomas Cooper, so that by the end of his reign, in 1820, the Society had become a stronghold of conservative values. Furthermore, as Paul Weindling has argued, the exemption of chartered societies from the Parliamentary Acts against seditious meetings strengthened the Royal Society's position over other scientific organizations.[40] According to the radical writer and publisher Richard Carlile, this deference to State control was actually embodied in systems that gave mankind a privileged position among God's creatures, and postulated that the natural world was made simply for man's convenience. In *An Address to Men of Science* (1821), Carlile put forward the view that depriving man of his dominion over nature would lead to 'the representative system of government', while pouring scorn on the 'priestly cosmologies' of those who 'openly countenanced systems of error and imposture because the institutions of the country were connected with them.'[41] The 'institutions' upheld were social as well as political.

With regard to women, conservative systematics accorded with female modesty in teaching them piety, chastity and submission at a time when alarm about the stability of the social order was coupled by reactionaries with an intolerance of atheism and a fear of female sexuality. Descriptive natural philosophy was beneficial to the sex whose religious devotion was thought to be impeded by superstitious tendencies, and whose imagination, which we have seen posed a predominantly sexual threat, was in need of regulation. In practical terms the religious corrective was sufficient, for, as Hannah More explained, 'When sanctified by Christianity, the imagination is a lion tamed.'[42] As to submission, if the natural order was divinely ordained, it followed by analogy that the social order was similarly sacred. Conservative systems therefore sanctified a social world in which women were traditionally represented as creatures at least one link lower than man in the 'great chain of being', which ascended from brute matter to Holy Spirit. This was a society which Burke, leading philosopher of the counter-revolution, delicately referred to as the 'age of chivalry'. And this was a society which the French insurrection had undermined. With characteristic nostalgia, Burke complained: 'Never, never more, shall we behold that generous loyalty to rank and sex, that dignified obedience, that subordination of the heart, which kept alive, even in servitude itself, the spirit of an exalted freedom.'[43] Hannah More and the Reverend Thomas Gisbourne, who fronted the popular revival of evangelicalism, agreed with Burke in celebrating liberty as a product of hierarchy. Humanity was subservient to God, as was matter to spirit, profanity to divinity, and woman to man. More and Gisbourne took pains to stress that woman's virtue was a function of her exclusion from the public sphere and woman's happiness a condition of her subjugation by man. Woman's morality, moreover, was traded off against a denial of her sexuality.[44]

Women to the right of the political spectrum and dissenters keen to exempt their creed from the charge of atheism sought to ratify the prevailing conservative hegemony by using their pens to justify the status quo. One way of so doing was to commend systematic natural philosophy. Margaret Bryan, about whom precious little is known save that she ran a school for girls and wrote on scientific subjects, is a prime candidate for examination. Bryan's first work, *A Compendious System of Astronomy* (1797), is a set of introductory lectures on optics, lenses, gravity, the atmosphere and magnets, as well as astronomy. In her first lecture, Bryan, echoing the systematists, confirmed that the student of nature was rewarded 'by the extention of ideas and the strength of judgement acquired; by which the human understanding is enabled to soar above vulgar prejudices, and to view the works of God with satisfaction, – deriving consolation from every object in nature.'[45] Warmed by the success of this venture, she published her *Lectures on Natural Philosophy* (1806) which embraced mechanics, pneumatics, acoustics, hydrostatics, magnetism, electricity and astronomy. Bryan, acknowledging a debt to Paley, revealed that these lectures were intended 'to induce a clear and enlarged conception of the profound wisdom, exquisite contrivance, and extensive benevolence of the Creator, in the formation, endowment, beauty and usefulness of his works.' Addressing her pupils, she stressed the ethical sobriety such knowledge of the Deity would yield by providing 'a defence against the vain sophistry of the world; arming you with a perpetual talisman, which . . . will secure you from all pernicious doctrines, and guard your religious and moral principles against all innovations.'[46] The counter-revolutionary tone of this passage is thinly veiled.

Advocating a conservative natural theology was expedient for women of a liberal persuasion, who, responding to public pressure, felt a need to distinguish their own liberality from that of the republicans. Naturally, dissent added urgency to the motive. The quaker Priscilla Wakefield, author of a number of scientific texts, paid little more than lip-service to dominant intellectual currents when she affirmed that the study of nature was 'the most familiar means of introducing suitable ideas of the attributes of the Divine being, by exemplifying them in the order and harmony of the visible creation'. There is scant further reference to the argument from design in *An Introduction to Botany* (1796); indeed, her use of natural theology appears to have been a ploy to circumvent censorship. In fact, Wakefield's main thrust is to present botany as a cure for the aristocratic malady – 'an antidote to levity and idleness'.[47] Like Edgeworth, Wakefield held that woman's prime virtue was usefulness. Both were stirred by the early commercial and political successes of an emergent industrial middle class, whose simple work ethic made the notion of active, but none the less feminine, womanhood an attractive proposition. In her contribution to the growing criticism of traditional education for women, *Reflections on the present condition of the female sex, with suggestions for its improvement* (1798), Wakefield maintained that 'There are many branches of science, in which women may employ their time and their talents, beneficially to themselves and the community, without destroying the peculiar characteristic of their sex, or exceeding the most exact limits of modesty and decorum.'[48] She went on to plead in Wollstonecraftian fashion for the better education of women with a view to making them rational creatures worthy of employment. She herself set the example, for as well as being a prolific author, she aided her merchant husband, and as an ardent philanthropist founded several frugality banks under the auspices of a Friendly Society that she established in 1798. The polemical content of Wakefield's scientific writings – her agitation for improved education for women and her reappraisal of what constituted female modesty – points to a distinction between science and the representation of science.[49] We will briefly survey the manner in which scientific discourse served women as a vehicle for conveying arguments about female cultural politics.

A reluctance to forego many of the ideals of womanliness appears to have been common to women regardless of political or religious persuasion. Women's domesticity, for instance, was rarely challenged. Indeed, Maria Edgeworth warned that women's 'imagination must not be raised above the taste for necessary occupations, or the numerous small, but not trifling pleasures of domestic life.'[50] The Reverend Thomas Gisbourne typified the consensus in placing the importance of the female character principally in three areas: making domestic life comforting to family and friends, forming and improving the manners and conduct of the other sex, and teaching children. Not even Wollstonecraft disputed these duties. One clue in understanding the proliferation of domestic ideologues among women lies in recognizing that within the domestic sphere, as by extension through pedagogy, women had a means of legitimately influencing others. Women's moral exertions in the private sphere were on trial in the 1790s, when the sex was relied upon to ward of the republicanism that threatened to disrupt the quietude of home and hearth. It became possible for women to be active patriots by guarding the home front, and this path to self-worth was one they would not have undermined; one woman reader of *The Anti-Jacobin* was quick to rebuke them for being overly concerned with public matters, 'as if the Jacobinism and Principles which you set up to oppose, did not disturb Domestic Felicity and Comfort as much as it does Kingdoms and Empires.'[51] Women leant on the authority that responsible motherhood and proud domesticity accorded to them. Margaret Bryan in her *Lectures on Natural Philosophy* celebrated her domestic internment, announcing, 'I rejoice in the titles of Parent and Preceptress.'[52] The frontispiece to her earlier work on astronomy (see plate 2) features Bryan, quill in hand, seated demurely in the comfort and retreat of her study beside her two daughters and surrounded by an array of scientific instruments. She acknowledges the role of motherhood in justifying public activity when she explains that 'The most tender claims upon my exertions, (the nature of which may be understood from the frontispiece) seconded by the encouragement of my friends' tempted her to publish. Elsewhere Bryan protested that she felt 'almost parental tenderness' for her pupils.[53] On balance, Bryan's homely respectability outweighed latent prejudices against the manly learnedness of scientific women and won her praise from the critics. Paradoxically then, women, by remaining within the

PLATE 2 Mrs Bryan and Children. Engraved by W. Nutter from a miniature by Samuel Shelley. The original painting was exhibited at the Royal Academy in 1797 and described mistakenly in the exhibition catalogue as 'A family, designed for a frontispiece to Mr. Bryan's lectures on astronomy.' Frontispiece to Margaret Bryan's *A Compendious System of Astronomy* (1797).

accepted bounds of femininity, were able simultaneously to extend them.

Once the shadow of revolution began to lift, domesticity, especially for 'serious Christians' and the self-conscious middle classes, was additionally seen, as Davidoff and Hall have shown, as a refuge from the sordid worldliness and corruption of the public sphere.[54] The rural domicile in particular, idealized time and time again in the novels of Jane Austen, was where simple virtues were best cultivated. There is some irony in the conservative middle classes adopting an essentially Rousseauist moral dualism, but in the nineteenth century the association of country life with the ignorance and barbarism of the peasant rabble, which in the 1790s was seen to possess the greatest potential for insurrection, was broken. Notions of rural tranquillity, innocence and Christian virtue accompanied movement of the wealthy middle classes out of industrializing cities and into suburbia. The domestic idyll grew to embrace home and garden, and botany became the most moral of sciences. The conservative trends which dominated cultural activity ensured that in academic botany advocates of natural systems were by the 1820s competing with proponents of Linnaean sexual classification. Women were among those who sought to erase sexuality from taxonomies of the plant kingdom. Already in 1801 Frances Arabella Rowden rendered Darwin's *Botanic Garden* respectable, as Ann Shteir points out, by 'changing his references to brides and connubial happiness to maternal love and suppression of sexual feeling'.[55] The natural historian Mary Roberts, backed the systematists with *The Wonders of the Vegetable Kingdom* (1822). But it was Jane Louden, writing in the 1830s and 40s, for whom 'there was something in the Linnaean system . . . excessively repugnant', and her landscape-gardener husband, John Claudius Louden, who presented the middle classes with a comprehensive blueprint of home-and-garden life based on the unlikely partnership of taste and practical utility. According to John Claudius Louden, 'Nothing contributes more to the moral and political government of the passions' than 'good taste'. While women had long been privileged as the premier arbiters of taste, Louden twisted the active nature of this role into a passive one, by identifying the serene, modest, beautiful, restrained and domiciled woman as being herself the model of tastefulness.[56] For Louden, domesticity was more than a sphere which contained women; it was femininity itself.

Domesticity was only one of a number of ideals which the majority of women writers on science either intentionally or unreflectively upheld.

Developmentally, women were generally considered retarded by comparison with men. Their lesser physical stature, it was argued, was certain indication that women, whether by divine intention or socio-medical recommendation, were not fitted for public station as were men.[57] By denying male protection, by being assertive or opinionated, women were seen to be transgressing the *physical* limits of their potential, and were condemned as non-women. The developmental argument was extended to embrace mind as well as body; women were thought to be somehow more primitive or infantile than men – a notion which pervaded thinking in disciplines like anatomy and, later, craniology – and it was unfeminine for women to prove otherwise.[58]

Returning to our women writers on science, we should note that they wrote for an audience of women and children; they would not have presumed to instruct men. Bryan wrote for 'those who have not studied mathematics'; Wakefield's works on botany took the form of 'familiar letters' or 'entertaining dialogues', even a 'catechism'. Another writer of scientific primers, Jane Marcet, *émigrée* daughter of a wealthy Swiss merchant and wife of the chemist Alexander Marcet, wrote 'conversations'. There were conversations on chemistry (1806) 'intended more especially for the Female Sex', political economy (1816), natural philosophy (1819) and vegetable physiology (1829). These writers affirmed their childlike benignity by confining their output to fragmentary, pedagogic and introductory texts; their tone was didactic, and their style narrative. They provided only fleeting glimpses of grand scientific schemes, partial accounts and circumscribed conclusions. There was no visionary Romanticism, no holistic, imaginative or Utopian view of knowledge. In short, these scientific texts by and for women contained nothing controversial that might open the possibility of reading political dissent of any kind, including, of course, republicanism. Maria Edgeworth, intending only compliment to the author, commented, 'Mrs. Marcet never goes one point beyond what she can vouch for truth and never practises any of the little arts of professed conversationalists to hide the *bounds* of knowledge or to excite sensation or surprise.'[59] Likewise, independent creativity was something these authors gladly denied to themselves. Bryan declared an awareness that her work 'will not procure me any honor on the score of originality', and Marcet insisted that 'successful perseverance' is only 'retrospectively ascribed to genius'.[60] These women assiduously avoided those controversial forms of knowing which smacked of the subjective epistemology and individualism of the

Romantics. Instead, they relied on attitudes to the acquirement, display and transference of knowledge that were consistent with the ideals of womanliness.

MARY SHELLEY AND MARY SOMERVILLE

The after-shock of the French Revolution was felt with little diminution in early nineteenth-century Britain, and fears of insurrection welled up whenever domestic unrest culminated in clamours for political reform. The years 1816–19 and 1727–30 were particularly troubled ones, during which time the political debates of the 1790s were relived. A case study from each of these periods will further illuminate the political dimensions of science and gender that were pivotal in orientating women's approach to understanding nature. First, we will consider Mary Shelley (daughter of Mary Wollstonecraft and William Godwin), whose topical novel *Frankenstein* (1818) elaborates both a powerful critique of Romantic science and an implicit rejection of conservative systematics. Finally, we will look at the scientific writings of Mary Somerville. Unlike most of the women writers on science we have so far considered, Mary Shelley and Mary Somerville achieved considerable fame – a celebrity status not unrelated to the fact that they both wrote as adults for adults.

Mary Shelley's *Frankenstein; or, The Modern Prometheus* tells the story of the egotistical Victor Frankenstein, who in the throes of a misguided and malignant passion creates a monstrous being, who, because deprived of parental nurture and rejected by society, becomes a 'fiend', turning on his master and destroying both him and all that he values. This, her most poignant novel, was Shelley's response to Lord Byron's challenge to write a ghost story. How was it that Shelley came to have the nightmarish vision from which her novel germinated, of 'the pale student of unhallowed arts kneeling beside the thing he had put together'?[61] From where did her conception derive of the 'unhallowed arts', which enabled Frankenstein to 'infuse a spark of being' (p. 101) into the creature, whose component parts he had collected from charnel-houses and graveyards?

Shelley herself provides the lead in an account of the origins of *Frankenstein*, which she prefixed to its 1831 edition. Here she recalled the summer of 1816 when she 'was a devout but nearly silent listener' to the philosophical discussions between Percy Shelley and Byron at the Villa

Diodati in Switzerland. The poets speculated on 'the nature of the principle of life' (p. 54), speaking of some experiments which Erasmus Darwin was rumoured to have conducted that induced a piece of vermicelli to wriggle of its own accord.[62] Shelley was thus led to muse: 'Perhaps a corpse would be reanimated; galvanism had given token of such things: perhaps the component parts of a creature might be manufactured, brought together, and endued with vital warmth.' (p. 54) Giovanni Aldini, Professor of Natural Philosophy at Bologna, attracted widespread popular notice to the theory of animal electricity in the first decade of the nineteenth century. He galvanized the corpse of the murderer Thomas Forster in 1803 producing a degree of muscular convulsion such 'as almost to give an appearance of re-animation'.[63] Aldini's experiments were imitated across Europe over the next few years. In all likelihood, Shelley was aware of these headlining developments in science. Similarly, she would have known of the conflicts in vitalist physiology — over whether activity was inherent in or superadded to matter — which followed in 1816 after the immanentist William Lawrence, physician to the Shelleys, published an attack on the transcendentalist theories of John Abernethy.[64]

Aside from the natural philosophers Shelley mentions by name, many critics have observed that the character of M. Waldman, the professor of chemistry who lures Frankenstein away from his attachment to Paracelsus and inspires his devotional feelings toward modern chemistry, bears a strong resemblance to Humphry Davy. Indeed, Shelley records in her journal for 28 October 1816: 'Read the Introduction of Sir H. Davy's 'Chemistry'; write.'[65] In a passage which recalls Davy's *Discourse*, Waldman derides the ancient chemists who 'promised impossibilities and performed nothing' in contrast to the moderns who promise little but,

> have indeed performed miracles. They penetrate into the recesses of nature and show how she works in her hiding-places. They ascend into the heavens; they have discovered how the blood circulates, and the nature of the air we breathe. They have acquired new and almost unlimited powers; they can command the thunders of heaven, mimic the earthquake, and even mock the invisible world with its own shadows. (p. 92)

Frankenstein's initiation into potent science has, moreover, much in common with Percy Shelley's own scientific experience. The poet's friend and biographer Thomas Jefferson Hogg, relating his first meeting with

Percy, tells how he was confronted with a passionate Davy-like lecture on the wonders of modern chemistry. An enduring interest in the subjective sciences of discovery, which had become aligned with the poetic imagination, surfaced in Percy's writings, notably in *Prometheus Unbound*.[66] Romantic philosophers unleashed their imaginations in natural philosophies which were unpredictable, unlimited and reactive; like Prometheus, they had stolen the fire of creativity from the gods and deployed it to their own ends. Lest we identify Frankenstein too closely with any individual actor, Shelley does not provide, as Christopher Small has indicated, any physical description of her natural philosopher: 'One might say that he has no physical body, he is all spirit and restless, inquiring mind.'[67] Victor Frankenstein is a convincing *mélange* of the day's leading proponents of Romantic science.

Frankenstein's fate is a lesson in the moral anarchy and potential destructiveness of Romantic science; his individual perception of reality – presented as a product of egocentricity not genius – becomes isolationist, leaving him ostracized from the community of mankind as he feverishly manufactures his monster. His passion for 'unhallowed arts' turns to obsession; ruining his health, causing him to neglect family and friends, and leading him alone into blind pursuit of the being to whom he had given life but not love, and who in return murdered his kin. Anne K. Mellor, in her perceptive study of Shelley's life and writings, draws attention to yet another and subtle way in which *Frankenstein* shows up subjective science as delusory: all the characters in the novel, except for the blind Father De Lacey, prejudge the monster by reading his character through his physiognomy, or outward appearance. Despite his protestations that social misery made him a wretch, the monster is persistently deemed 'evil' because he is 'hideous'.[68] Frankenstein's fate is determined by his passion for an irresponsible science, which prevents him (and his monster) from taking a place in society, and more particularly from fulfilling his domestic duties.

Shelley's linking of creative science to an irresponsible and destructive masculinity is perhaps best understood in relation to her novel's autobiographical symbolism. Frankenstein's failure to sustain any domestic happiness can be seen to reflect Shelley's frustration with Percy's *modus vivendi* – her reluctance to accept the Victor in him.[69] Percy was in a crucial sense not the husband Mary Shelley wanted and needed. His Romantic sensibility demanded constant female appreciation, and when sympathy was not forthcoming from his wife, he sought it elsewhere; in

Claire Clairmont – Mary's stepsister who lived with the young couple for several years – and later in Emilia Viviani and Jane Williams. Before she wrote *Frankenstein*, Shelley had already lost a child, whose death Percy barely mourned, and she was pregnant while she was writing the novel. How could she be certain of a secure connubial future, when at the end of 1816 Percy's first wife, Harriet, was driven tragically to commit suicide? With respect to an identification of Frankenstein with William Godwin, Shelley's empathy with the monster takes on significance as an expression of anger at her own father's neglect of her as a child. In this light, the novel's chilling murder of her father's and her son's namesake, 'little William', is an outburst of tortured feelings of both patricide and infanticide. The symbolism becomes even more convoluted when we note that William was the name chosen for the boy the Godwins were convinced Mary was going to be, and the boy whose existence was realized by Godwin's second wife, Mary Jane Clairmont.[70]

Seeing Frankenstein as a latter-day Godwinian throws light on Shelley's political critique of Romantic idealism. *Frankenstein* was composed during the post-war slump, a time which witnessed a short-lived Jacobin revival – the literary tastes of which were catered for by Richard Carlile and William Hone, who, risking prosecution, reprinted Paine and the French classics. We know that Mary Shelley was well versed in this literature, but whether she subscribed to its idealism is more doubtful; she was later to state in her journal: 'since I had lost Shelley I have no wish to ally myself to the Radicals – they are full of repulsion to me – violent without any sense of Justice – selfish in the extreme – talking without knowledge – rude, envious and insolent – I wish to have nothing to do with them.'[71] While we would be too hasty in branding Shelley's timely novel as reactionary, she does invoke a standard conservative trope in her use of monster metaphor. Counter-revolutionaries in the 1790s symbolized the violent anarchy that resulted when the revolutionary mob latched on to ideas of reform by depicting Utopian reformers as breeders of monsters who threaten to destroy their creators. Lee Sterrenburg has suggested that Shelley, internalizing the politics, made Frankenstein's monster symbolize the domestic carnage wrought by Jacobinism. This imagery had personal significance for Shelley as the daughter of two radicals who had oft been charged with monstrosity. Godwin hoped for the regeneration of society through the creation of a benevolent, rational and free new race – a motherless race produced through social engineering.[72] Victor Frankenstein, whose idealistic aspirations are selfish, misguided and ultimately destructive, and whose monster degenerates into a revengeful fiend, is a cruel parody of Godwin and Godwinian social reform.

While the monster metaphor reveals a reactionary dimension to her novel, Shelley's attack on illuminist individualism embodies a cautionary address to conservative moralists. The Abbé Barruel – whom the Shelleys had read – claimed for his conspiracy theory of the French Revolution, that 'the Illuminating Code . . . engendered that disastrous monster called Jacobin.'[73] It is clear that Shelley intended to link Frankenstein with the 'illuminati' by locating his initiation into 'unhallowed arts' at Ingolstadt, birthplace of Adam Weishaupt's revolutionary sect. Although *Frankenstein* is in many ways a late contribution to the political debates of the 1790s, it is no less a commentary on post-war Britain, in that the enlightened individualism, which had inspired the Girondists and Jacobins, had been redefined by a group with entirely different social aims – the evangelicals.

By the early nineteenth century, evangelicalism had gained disciples across denominations. Its central theme was that mankind, though tainted with original sin, could obtain salvation and an eternal afterlife through an awakening to revealed religion. The emphasis was on individual conversion or rebirth. Evangelicalism was especially compatible with the role of women sanctioned by a conservative use of systematics, and women, long renowned for religious enthusiasm, were prominent among its proselytes.[74] Active in spreading the word was Elizabeth Hamilton, who, while recommending bringing ardour to the study of religion, protested against a rational Christianity that appealed to the understanding: 'Religion is then to be learned as a science, a mere matter of speculation; it is to be propounded to the unbiased judgement as an object of curiosity, almost as worthy of investigation as the laws of electricity or magnetism.' Hamilton maintained that 'religious sentiment' should instead be 'blended with all that touches the heart and charms the imagination.'[75] With its celebration of the individual imagination, enthusiasm was but illuminism by another name. It is no accident that Shelley employed religious terminology to describe Frankenstein's passion for his loathsome creativity; she referred to his 'enthusiastic frenzy' and 'enthusiastic madness'. For Shelley all ideologues were blind to social reality.

As a critic of Romantic science, Shelley broke with convention, for she did not endorse systematic natural theology. Frankenstein defied not respected the natural order when he transformed dead matter into a living being and produced not an Adam but a monster. Moreover, the sublime natural environments Shelley chose as the setting for several scenes totally failed to inspire either Frankenstein or his monster with feelings of godliness. In stark contrast, the Alpine valleys of Chamonix, the rocky, barren Orkneys, and the magnificent polar ice-caps formed the backdrop for unforgiving confrontations between these two despairing characters. It can be argued that it was Frankenstein's inability to appreciate a systematic world view which led to his fall, yet Shelley points to the inadequacy of nature alone, even in its most stupendous forms, to engender morality. Shelley again played with extremes as she mocked the traditional social order for which systematics provided an ideological prop. Outraged by a male appropriation of creativity, she fictionalized Frankenstein's powers to a wry perfection in endowing him with the ability to reproduce without women. Shelley registered her objection to the popular theory of sexual complementarity — which defined male and female ideals as opposite and complementary — by working deftly with stereotypes. She contrasted Frankenstein's Promethean role of plasticator with a portrayal of her female characters as essentially passive, or in the apt wording of Darwin — 'pliant'.[76] Elizabeth Lavenza, Caroline Frankenstein and Justine Moritz are all carers; self-sacrificing women, whose tame domesticity, and misplaced dependence upon men who fail to look after them, prevents them saving their own lives. The novel's only image of a self-determining, independent woman is drawn through Frankenstein's doubtful meditations on creating a female companion for his monster; and this image is shattered before it materializes, as Frankenstein violently tears apart his Eve.[77] Like her mother, Shelley idealized the egalitarian family unit as a regenerating force in society. Her own desire for equality was woefully unfulfilled in her marriage, and we can see her writing of *Frankenstein* both as an assertion of her independence and of her adulthood.

A close reading of *Frankenstein* dramatically highlights the degree of confluence which persisted between languages of science and sexuality well into the nineteenth century. If we turn to a consideration of Mary Somerville, who ventured into print with heavyweight texts on astronomy and the physical sciences in the 1830s, we will see that despite her conscious attempts to objectivize science and break with the genre of primers, she was not able to disguise gender interests.

Somerville was a legitimate voice of the scientific elite, a protégé of John Playfair, and a valued outpost of the 'Cambridge network', which claimed responsibility for importing French mathematical analysis and thus arresting the decline of science at home.[78] The men of science who fostered an interest in Somerville's scientific aspirations – among them, Playfair, John Herschel, Charles Babbage and William Whewell – were aware that lingering suspicions of things French might frustrate their attempts to popularize mathematical analysis. Thus it became imperative for them to move science off the battleground where political, religious and class conflicts were fought. In 1831, Whewell and Babbage were instrumental in founding the British Association for the Advancement of Science, whose core members pioneered painstaking endeavours to denude scientific discourse of overtly partisan political and religious content. On both counts they aimed for the largely uncontroversial middle ground with common-sense conservatism wedded to liberal Anglicanism.[79] In their conscious striving for an objective scientific practice, the scientific elite side-stepped Paleyean systems. Somerville, recalling such efforts in her memoirs, claimed that the 'purely mental conceptions of numerical and mathematical science which have been by degrees vouchsafed to man . . . must have existed in that sublimely omniscient Mind from eternity.'[80] Once mathematical language was seen as divine, the moral high ground could be easily claimed without making overtures to natural theology. Scientific discourse was further stripped of the ubiquitous language of the passions. As in systems, science was placed above the subjective knowledge gained through direct experience of natural phenomena; it became, as Somerville confirmed in her début *opus*, *The Mechanism of the Heavens* (1831) – a translation of *La Place* – 'the pursuit of truth, which can only be attained by patient and unprejudiced investigation.'[81]

The image of Somerville as an enlightened Frenchified thinker recalls Henry De la Beche's 1832 lithograph, 'The light of science dispelling the darkness which covered the world'. In it, science is personified as a woman, complete with frills and French bonnet. She floats delicately on a cloud from where she shines her lamplight on to a globe, rolling back the clouds encircling it.[82] After publishing her *Mechanism*, Somerville was seen by most of her contemporaries in such near-heroic light. Her viability in the scientific culture of her day owed much to her ability to anchor what Jane Marcet called her talents of 'masculine magnitude' in the elitist ideology cultivated by the gentlemen of science. Sustaining an identity of vision with her peers, however, involved Somerville in the practice of what should by now be a familar gender duality. The scientific journalist David Brewster reflected general opinion when he described her as 'a Mathematician of the very first rank, with the gentleness of a woman, and all the simplicity of a child.'[83] We have already seen how it benefited women in public activity to comply with the ideals of womanliness, but happily for historians, Mary Somerville's manuscript correspondence opens the door to a rather different image.

Somerville possessed a relentless ambition to succeed. When Herschel failed to confirm the results of her experiments to establish the magnetizing properties of ultra-violet light, which appeared in the *Philosophical Transactions* (1826), she burnt all her copies of the paper. This act, confessed in the first draft of her autobiography, was edited out of the posthumously published version by her daughter Martha. Martha chose to present her mother as 'always diffident about her writings'.[84] Fiercely competitive with her scientific friends, Somerville wanted desperately to be seen as a topical and independent author. On completing *On the Connexion of the Physical Sciences* (1834), hot on the heels of William Whewell's *Bridgewater Treatise*, she wrote with relief to her son: 'I am glad my book is in the press at last and am still happier to find Whewell's differs too much from mine to knock out my brains.' Two months later she complained to him, 'I see also J. Herschel's *Astronomy* is out which will do mine no good.'[85] These are hardly the sentiments of a simple child. In contrast to the previous generation of women writers on science, it was with reluctance, not acquiescence, that Somerville concluded that women were not innovative. Reflecting on her long life in science, she wrote in the first draft of her memoirs:

> I was conscious that I had never made a discovery myself, that I had no originality. I have perseverance and intelligence but no genius, that spark from heaven is not granted to the sex, we are of the earth, earthy, whether higher powers may be allotted to us in another state of existence God knows, original genius in science at least is hopeless in this.[86]

The feelings Somerville expressed in manuscript – ambition, determination and competitiveness – point to a general problem concerning female authorship, namely, the separation of public from private self. The commonest strategy employed by women writers on science during this period, in their attempts to reconcile public ideals of womanliness with their personal gender identity, was to remain within the traditional bounds of pedagogic discourse. Women like Bryan, Wakefield and Edgeworth adapted languages of science to fit such discourse; they deployed conversational or didactic narratives set more often than not in the class-room; they wrote primers and addressed children and women. Moreover, they wrote about sciences which affirmed conventional gender roles through an appeal to Providence. Women speaking their mind through public discourse, like Wollstonecraft, could be seen to be overstepping the boundaries of the domestic sphere, which encompassed the roles of wife, mother and teacher, and thereby risked being charged with vulgarity and immorality. Any assertion of personhood, such as was widely understood to underlie advocacy of Romantic science, could not be accommodated within domestic ideologies which anchored female virtue to self-denial.

But for Mary Shelley and Mary Somerville, who targeted an adult audience of men and women, a recourse to pedagogic techniques was not available. Mary Poovey has suggested that Shelley dealt with her ambivalence towards female self-assertion by distancing herself from her narrative. In splitting the narrative of *Frankenstein* between three male characters, Shelley averted the charge she most feared: that of being labelled, as her mother had been, a monstrous female artist.[87] Somerville similarly distanced herself from her scientific writings, striving for the elimination of her identity from her work. In fact, John Herschel, reviewing *On the Mechanism of the Heavens*, praised the book for this very feature. Remarking on the absence of 'anything like female vanity or affectation', he mused, 'She seems entirely to have lost sight of herself.'[88] This self-denial, or masking of personal gender identity was the price women had to pay for entering public discourse, especially male-dominated discourses like science. But there were gains to compensate for such concessions, since, as we have seen, women writing about science were at the same time dealing fundamentally with female morality, their own, and that of women in general. In this sense, women writers on science were able to explore the continuity between domestic and political life – an issue characteristic of the age of revolution.

NOTES

1 Olivia Smith, *The Politics of Language 1791–1819* (Clarendon Press, Oxford, 1984).
2 For the historical approach to the meanings of single terms see Raymond Williams, *Keywords: A Vocabulary of Culture and Society* (Penguin, London

1983); S. I. Tucker, *Protean Shape, A Study in Eighteenth-Century Vocabulary and Usage* (Athlone Press, University of London, 1967).

3 Sandra Harding, *The Science Question in Feminism* (Open University Press, 1986); Elizabeth Fee, 'Critiques of modern science; the relationship of feminism to other radical epistemologies' in Ruth Bleier (ed.), *Feminist Approaches to Science* (Pergamon Press, Oxford, 1986); Alison Kelly, 'The construction of masculine science', *British Journal for the Sociology of Education*, 6 (1985), pp. 133–54.

4 Sylvana Tomaselli, 'The Enlightenment debate on women', *History Workshop Journal*, 20 (1985), pp. 101–24; Joan Landes, 'Women and the public sphere: a modern perspective', *Social Analysis*, 15 (1984), pp. 20–31; Joan Scott, 'Women in history', *Past and Present*, 101 (1983), pp. 141–57; Elizabeth Fox-Genovese, 'Placing women's history in history', *New Left Review*, 133 (1982), pp. 5–29.

5 Mary Wollstonecraft, *A Vindication of the Rights of Woman*, 2nd edn (1792), ed. with intro. Miriam Brody Kramnick (Penguin, 1985), p. 91.

6 A classic article is A. O. Lovejoy, 'On the discrimination of Romanticisms', *Publications of the Modern Language Association of America*, 39 (1924), pp. 229–53. See also Roy Porter and Mikuláš Teich (eds), *Romanticism in National Context* (Cambridge University Press, Cambridge and New York, 1988).

7 Marilyn Butler, *Romantics, Rebels and Reactionaries; English Literature and its Background 1760–1830* (Oxford University Press, Oxford, 1981); David Knight, 'The physical sciences and the Romantic Movement', *History of Science*, 9 (1970), pp. 54–75.

8 David Knight, 'The scientist as sage', *Studies in Romanticism*, 6 (1967), pp. 65–88; Walter Riesse, 'Romantic natural philosophy and experimental method', *Studies in Romanticism*, 2 (1962), pp. 12–22.

9 On organization see Karl M. Figlio, 'The metaphor of organisation: an historiographical perspective on the bio-medical sciences of the early nineteenth century', *History of Science*, 14 (1976), pp. 17–53 and 'Theories of perception and the physiology of mind in the late eighteenth century', *History of Science*, 13 (1975), pp. 171–212. On sensibility see Christopher Lawrence, 'The nervous system and society in the Scottish Enlightenment' in Barry Barnes and Steven Shapin (eds), *Natural Order* (Sage Publications, Beverley Hills, California, 1979), pp. 19–39; Simon Schaffer, 'Genius in romantic natural philosophy' in Andrew Cunningham and Nicholas Jardine (eds), *Romanticism and the Sciences* (Cambridge University Press, Cambridge, 1990), ch. 6.

10 W. Cameron Walker, 'Animal electricity before Galvani', *Annals of Science*, 2 (1937), pp. 84–113.

11 Nora Cook and Derek Guiton, *Shelley's Venomed Melody* (Cambridge

University Press, Cambridge, 1986), pp. 23—4. See also A. Hayter, *Opium and the Romantic Imagination* (Faber and Faber, London, 1968).

12 Humphry Davy, *Researches Chemical and Philosophical Chiefly Concerning Nitrous Oxide* (1800), reprinted in *Collected Works*, ed. J. Davy (9 vols, London, 1839—40), vol. 3, pp. 289—90: cited by Knight, 'The scientist as sage' in relation to Davy's *Consolations in Travel* and by Schaffer in 'Genius and romantic natural philosophy' in relation to auto-experimentation.

13 Knight, 'The physical sciences and the romantic movement'.

14 Humphry Davy, *A Discourse, Introductory to a Course of lectures on Chemistry* (London, 1802), p. 17.

15 For a critical assessment of Davy's relation to the Lake poets see Trevor H. Levere, 'Humphry Davy, "The sons of genius", and the idea of glory' in Sophie Forgan (ed.), *Science and the Sons of Genius: Studies on Humphry Davy* (Science Reviews, London, 1980), pp. 33—58, and *Poetry Realised in Nature: Samuel Taylor Coleridge and Early Nineteenth-Century Science* (Cambridge University Press, Cambridge, 1981); Roger Sharrock, 'The chemist and the poet: Sir Humphry Davy and the preface to the *Lyrical Ballads*', *Notes and Records of the Royal Society of London*, 17 (1962), pp. 57—76.

16 Humphry Davy, *Discourse*, p. 16.

17 Robert Darnton, *Mesmerism and the End of the Enlightenment in France* (Harvard University Press, Cambridge, Mass., 1968). See also Roy Porter, 'Under the influence: mesmerism in England', *History Today*, 35 (1985), pp. 22—9 and Roger Cooter, 'The History of Mesmerism in Britain: poverty and promise' in Heinz Schott (ed.), *Franz Anton Mesmer und die Geschichte des Mesmerismus* (Franz Steiner Verlag Wiesbaden GMBH, Stuttgart, 1985), pp. 152—62.

18 Geoffrey Sutton, 'Electric medicine and mesmerism', *Isis*, 72 (1981), pp. 375—92.

19 George Bloch, *Mesmerism: A Translation of the Original Scientific and Medical Writings of F. A. Mesmer*, intro. E. R. Hilgard (William Kaufmann, California, 1980), p. 54.

20 Humphry Davy, *Discourse*, p. 10.

21 Darnton, *Mesmerism*, ch. 5.

22 John Graham, 'Lavater's physiognomy in England', *Journal of the History of Ideas*, 22 (1961), pp. 561—72.

23 John Martin, *Animal Magnetism Examined: in a Letter to a Country Gentleman* (London, 1790), p. 20.

24 Letter to Horace Walpole, Sept. 1788, in William Roberts (ed.), *Memoirs of the Life and Correspondence of Mrs Hannah More* (4 vols, London, 1834), vol. 2, p. 120.

25 Hannah More, *Strictures on the Modern System of Female Education* (2 vols, London, 1799), vol. 1, p. 78.

26 G. S. Rousseau, 'Nymphomania, Bienville and the rise of erotic sensibility' in Paul-Gabriel Bounce (ed.), *Sexuality in Eighteenth-century Britain* (Manchester University Press, and Barnes and Noble, Manchester and New Jersey, 1982), pp. 95–119. See also the essays in Roy Porter and G. S. Rousseau (eds), *Sexual Underworlds of the Enlightenment* (Manchester University Press, Manchester, 1987).

27 This thesis is well argued in David Morse, *Romanticism, a Structural Analysis* (Macmillan, Barnes and Noble, London and New Jersey, 1982) and in Maurice Hindle's introduction to *Caleb Williams* (Penguin, Harmondsworth, 1988).

28 Sharrock, 'Lavater's physiognomy in England'.

29 Cynthia Griffin Wolf, 'The Radcliffean Gothic model: a form for feminine sexuality', *Modern Language Studies*, 9 (1979), pp. 98–113. For background on Radcliffe see John Garret, *Gothic Strains and Bourgeois Sentiments in the Novels of Mrs. Ann Radcliffe and her Imitators* (Arno Press, New York, 1980).

30 David Punter, '1789: the sex of revolution', *Criticism*, 24 (1982), pp. 201–17.

31 R. M. Janes, 'On the reception of Mary Wollstonecraft's *A Vindication of the Rights of Woman*', *Journal of the History of Ideas*, 39 (1978), pp. 293–302. Barbara Taylor deals with Wollstonecraft's audience in *Eve and the New Jerusalem* (Virago, London, 1983), ch. 1.

32 For discussion of Darwin and Romanticism see Maureen McNeil, *Under the Banner of Science: Erasmus Darwin and His Age*, (Manchester University Press, Manchester, 1987); Desmond King-Hele, *Erasmus Darwin and the Romantic Poets* (Macmillan, London, 1986).

33 Reverend Richard Polwhele, *The Unsex'd Females* (London, 1798, New York, 1800): quotations from pp. 4, 21 and 9. For detail on reception of *The Botanic Garden* see Norton Garfinkle, 'Science and religion in England 1790–1800', *Journal of the History of Ideas*, 16 (1955), pp. 376–88.

34 Maria Edgeworth, *Letters for Literary Ladies* (London, 1795), p. 66.

35 Mary Wollstonecraft, *Vindication*, p. 303.

36 The *Quarterly Review*, 10 (1814), saw fit to remark of *Patronage* that Edgeworth was apt to give 'all her characters a tincture of science, and to make them fond of chemistry and mechanics.' For background on Edgeworth see Marilyn Butler, *Maria Edgeworth: A Literary Biography* (Clarendon Press, Oxford, 1972).

37 For background on natural theology see C. C. Gillispie, *Genesis and Geology* (Harvard University Press, Cambridge, Mass., 1951); John Gascoigne, 'From Bentley to the Victorians: the rise and fall of British Newtonian natural theology', *Science in Context*, 2 (1988), pp. 219–56.

38 Simon Schaffer, 'Natural philosophy and public spectacle', *History of Science*, 21 (1983), pp. 1–43, at p. 32.

39 John Robison, *Proofs of a conspiracy against all the religions and governments of Europe, carried on in the secret meetings of free masons, illuminati, and reading societies* (London, 1797). For discussion see J. B. Morrell, 'Professors Robison and Playfair, and the Theophobia Gallica: natural philosophy, religion and politics in Edinburgh, 1790–1815', *Notes and Records of the Royal Society of London*, 26 (1971), pp. 43–63.

40 Paul Weindling, 'Science and Sedition: how effective were the acts licensing lectures and meetings, 1795–1819?' *British Journal for the History of Science*, 13 (1980), pp. 139–53.

41 Richard Carlile, *An Address to Men of Science* (London, 1821).

42 Hannah More, *Strictures*, p. 237.

43 Edmund Burke, *Reflections on the Revolution in France* (London, 1790), p. 113.

44 Hannah More, *Strictures*; Reverend Thomas Gisbourne, *An Enquiry into the Duties of the Female Sex* (London, 1797). See also Nancy Cott, 'Passionless: an interpretation of Victorian sexual ideology, 1790–1850', *Signs*, 4 (1978), pp. 219–36.

45 Margaret Bryan, *A Compendious System of Astronomy* (London, 1797), preface and p. 4; *DNB*, eds. Leslie Stephen and Sidney Lee (London 1808–9), vol. 3, p. 154.

46 Margaret Bryan, *Lectures on Natural Philosophy* (London, 1806), at p. 1 and 'Address'.

47 Priscilla Wakefield, *An Introduction to Botany* (London, 1796), preface; *DNB*, vol. 20, pp. 455–6.

48 Priscilla Wakefield, *Reflections on the present condition of the female sex, with suggestions for its improvement* (London, 1798), pp. 8–9.

49 For an excellent interpretation of scientific representation in such works see Greg Meyers, 'Science for women and children: the dialogue of popular science in the nineteenth century', in John Christie and Sally Shuttleworth (eds), *Nature Transfigured, Science and Literature, 1700–1900* (Manchester University Press, Manchester and New York, 1989). See also J. McDermind, 'Conservative feminism and female education in the eighteenth century', *History of Education*, 18 (1989), pp. 309–22.

50 Maria Edgeworth and Richard Lovell Edgeworth, *Practical Education* (London 1798), pp. 550. The Edgeworths also inform their readers that they have substituted the words 'domestic happiness' for the phrase 'success in the world', p. 725.

51 *The Anti-Jacobin*, vol. for 1797–8; Letter from a Letitia Sourby, 18 Dec. 1797, p. 44.

52 Margaret Bryan, *Natural Philosophy*, 'Address'.

53 Margaret Bryan, *Astronomy*, pp. viii, iv.

54 Leonore Davidoff and Catherine Hall, *Family Fortunes* (Hutchinson, London, 1987).

55 Ann Shteir, 'Linnaeus's daughters: women and British botany' in Barbara J. Harris and John McNamara (eds), *Women and the Structure of Society: Selected Research from the Fifth Berkshire Conference on the History of Women* (Duke University Press, Durham NC, 1984).

56 Jane Louden, *Botany for Ladies* (London, 1842), p. iii. See Bea Howe, *Lady with Green Fingers: the Life of Jane Louden* (Country Life, London, 1961); John Louden, *A Treatise on Forming, Improving and Managing Country Residences and the Choice of Situation Appropriate to Every Class of Purchaser*, vol. 1, p. 45: cited by Davidoff and Hall, *Family Fortunes*, p. 191.

57 The text which most notoriously sparked such thinking was Jean-Jacques Rousseau's, *Émile* (1762), ed. with intro. P. D. Jimack (Dent, London and Melbourne, 1984).

58 Elizabeth Fee, 'Nineteenth-century craniology: the study of the female skull', *Bulletin of the History of Medicine*, 53 (1979), pp. 415–33.

59 Letter to Harriet Beaufort, 5 Jan. 1822, in Christina Colvin (ed.), *Maria Edgeworth, Letters from England 1813–1844* (Clarendon Press, Oxford, 1971), p. 308.

60 Margaret Bryan, *Astronomy*, preface, p. vii; Jane Marcet, *Bertha's Visit to her Uncle in England* (3 vols, London, 1830), vol. 1, p. 268.

61 Mary Shelley, *Frankenstein, or the Modern Prometheus* (1831), reprinted with an excellent introduction by Maurice Hindle (Penguin, Harmondsworth, 1985). In 1831, Shelley for the first time provided an introduction to her novel. This edition will hereafter be cited by page number.

62 In his 1818 preface to *Frankenstein*, Percy Shelley also referred to Darwin as he whet the appetite of his wife's readers: 'The event on which this fiction is founded has been supposed by Dr Darwin and some of the physiological writers of Germany, as not of impossible occurrence.' Ibid., p. 57.

63 Giovanni Aldini, *An Account of the Late Improvements in Galvanism* (London, 1803), p. 194. See Paul Fleury Mottelay, *Bibliographical History of Electricity and Magnetism* (C. Griffin, London, 1922), pp. 305–7.

64 Theories of Immanance proposed that activity was innate in matter, while theories of Transcendence supposed activity was somehow superadded to matter. See L. S. Jacyna, 'Immanance or transcendence; theories of life and organisation in Britain 1790–1835', *Isis*, 74 (1983), pp. 311–29; Owsei Temkin, 'Basic science, medicine, and the Romantic era', *Bulletin of the History of Medicine*, 37 (1963), pp. 97–129.

65 *Mary Shelley's Journal*, ed. Fredrick L. Jones (University of Oklahoma Press, 1947), p. 67. Laura Crouch has argued persuasively that it was Davy's *Discourse* (1802) rather than his *Elements of Chemical Philosophy* (1812) that Mary was reading in 1816: 'Davy's *A Discourse, Introductory to a*

course of Lectures on Chemistry: a possible source of *Frankenstein', Keats–Shelley Journal*, 27 (1978), pp. 35–44.

66 For Frankenstein's resemblance to Percy Shelley see William Veeder, *Mary Shelley and Frankenstein – the Fate of Androgeny* (University of Chicago Press, 1986); Christopher Small, *Ariel like a Harpy, Shelley, Mary and Frankenstein* (Gollancz, London, 1972); P. D. Fleck, 'Mary Shelley's Notes to Shelley's Poems and Frankenstein', *Studies in Romanticism*, 6 (1967), pp. 226–54. For Percy's scientism see T. J. Hogg, *The Life of Percy Bysshe Shelley* (repr. 2 vols, London, 1933). Hogg quotes Shelley's scientific assault in 1810 thus: 'There are many mysterious powers, many irresistible agents. What a mighty instrument would electricity be in the hands of him who knew how to wield it, in what manner to direct its omnipotent energies; and we may command an indefinite quantity of the fluid: by means of electrical kites we may draw down the lightning from heaven! What a terrible organ would the supernal shock prove, if we were able to guide it; how many of the secrets of nature would such a stupendous force unlock! The galvanic battery is a new engine; it has been used hitherto to an insignificant extent, yet it has wrought wonders already' (vol. 1, p. 51). For the scientific imagery in Percy's own version of the Promethean myth see Carl Grabo, *A Newton Among Poets – Shelley's Use of Science in Prometheus Unbound* (University of North Carolina Press, Chapel Hill, 1930).

67 Small, *Ariel like a Harpy*, p. 102.

68 Ann K. Mellor, *Mary Shelley, her Life her Fiction her Monsters* (Routledge, New York and London, 1988), ch. 7.

69 Early in his career Percy adopted the pen name 'Victor'. Percy's sympathy for Frankenstein also surfaces in his posthumously published review of *Frankenstein* in *The Atheneum*, no. 263 (1832), p. 730. The review was written in 1818.

70 U. C. Knoepflmacher, 'Thoughts on the aggression of daughters' in George Levine and U. C. Knoepflmacher (eds), *The Endurance of Frankenstein, Essays on Mary Shelley's Novel* (University of California Press, 1979), pp. 88–119.

71 Mary Shelley's Journal, p. 205 entry for 1838.

72 Lee Sterrenburg, 'Mary Shelley's monster: politics and psyche in Frankenstein' in Levine and Knoepflmacher (eds), *The Endurance of Frankenstein*, pp. 143–71.

73 Quoted by Sterrenburg, ibid., p. 156.

74 Jane Rendall, *The Origins of Modern Feminism: Women in Britain, France and the United States, 1780–1860* (Macmillan, London, 1985); Doreen Rosman, *Evangelicals and Culture* (Croom Helm, London and Canberra, 1984); Davidoff and Hall, *Family Fortunes*.

75 Elizabeth Hamilton, *Letter on the Elementary Principles of Education* (2 vols, London, 1801), vol. 1, pp. 112, 129.

76 Erasmus Darwin, *A Plan for the Conduct of Female Education in Boarding Schools* (Derby, 1797), p. 10. Despite his reformist rhetoric, Darwin held conventional prescriptive views on the female character.

77 Mellor, *Mary Shelley*, ch. 6.

78 David Philip Miller, 'The revival of the physical sciences in Britain, 1815–1840', *Osiris*, 2nd series, 2 (1986), pp. 107–34; P. M. Harman (ed.), *Wranglers and Physicists, Studies on Cambridge Mathematical Physics in the Nineteenth Century* (Manchester University Press, Manchester and New Hampshire, 1985); Jack Morrell and Arnold Thackray, *Gentlemen of Science, early Years of the British Association for the Advancement of Science* (Clarendon Press, Oxford, 1981); Harvey W. Becher, 'William Whewell and Cambridge Mathematics', *Historical Studies in the Physical Sciences*, 11 (1980), pp. 1–48; Susan Faye Cannon, *Science in Culture: The Early Victorian Period* (Dawson and Science History Publications, New York, 1978).

79 Morell and Thackray, *Gentlemen of Science*, pp. 225–9.

80 Martha Somerville, *Personal Recollections, from Early Life to Old Age of Mary Somerville* (London, 1873), pp. 140–1.

81 Mary Somerville, *The Mechanism of the Heavens* (London, 1831), p. 2. This work was originally commissioned for Lord Brougham's Society for the Diffusion of Useful Knowledge: see J. N. Hayes, 'Science and Brougham's Society', *Annals of Science*, 20 (1964), pp. 227–41.

82 Paul J. McCartney, *Henry De La Beche: Observations of an Observer* (Friends of the National Museums of Wales, Cardiff, 1977): see pp. 54–5.

83 Brewster to J. D. Forbes, Sept. 1829: quoted in Elizabeth Patterson, *Mary Somerville and the Cultivation of Science 1815–1840* (Martinus Nijhoff Publishers, The Hague, 1983), p. 53.

84 Somerville, *Personal Recollections*, p. 120.

85 The Somerville Collection at the Bodleian Library, Oxford. Dep.c.361 MSIF–1, letter to Woronzow Greig, 24 Apr. 1833. Letter to Woronzow Greig, 14 June 1833.

86 Dep.c.355 MSAU–2, first draft of autobiography, refoliated in Jan. 1989 by Dr Judith Priestmen, p. 168.

87 Mary Poovey, 'My hideous progeny: Mary Shelley and the feminization of Romanticism', *PMLA*, 95 (1980), pp. 332–47.

88 John Herschel, review of *Mechanism of the Heavens*, *Quarterly Review*, 47 (1832), p. 548.

2

The Laws of Health:
Women, Medicine and Sanitary Reform, 1850–1890

Perry Williams

Victorian sanitary reform usually means to us parliamentary commissions and Public Health Acts, sewer building and water supplies: that is to say, matters of government administration and engineering, all very much the work of men. But there was another side to sanitary reform, which women claimed for themselves. As Susan Rugeley Powers put it in a tract of 1859:

> The great field of sanitary labor may be divided into two parts: the amelioration of injurious external circumstances, and the reform of injurious habits and customs. Of these parts the former belongs principally to man, the latter principally to woman. It is for man's comprehensive mind to devise schemes for draining and cleansing our towns, for improving dwellings, and for placing the necessaries of life within the reach of all; and it is for his strong hand to execute these schemes. It is for him to discover the laws of health and to teach and apply them where he can. It is for woman, in her functions of mother, housewife, and teacher, to effect those urgently needed changes in infant management, domestic economy, education, and the general habits of her own sex, without which humanity could never attain to its destined state of bodily perfection, though all injurious external circumstances were changed. It is for her to teach and apply the laws of health in her own province, where man cannot act.[1]

Powers was the secretary of the Ladies' National Association for the Diffusion of Sanitary Knowledge, which was founded in 1857 – the first sanitary organization to have a national scope.[2] The Ladies' Sanitary

Association, as it soon became known, enjoyed widespread and prestigious support amongst women of the middle and upper classes, and it continued in existence up to the end of the century as the central umbrella organization for what Powers called 'woman's work in sanitary reform': the distinctive and essential work of health education which it needed women to do.

My original brief for this chapter was to write on women as medical practitioners, and to focus on women sanitary reformers may seem an odd way of fulfilling it. Sanitary reform is not what first springs to mind when one thinks of medical practice. Indeed, it is probably the *last* thing to spring to mind; being preventative medicine, without an institutional base, and practised by lay people, it is about as remote as possible from the professional, hospital-based, curative medicine which has become our paradigm of health care. However, I believe that it is important to maintain a broad sense of what constitutes medical practice, especially when studying women practitioners. In medicine, as in science, as in every form of professional or occupational hierarchy, there is gender stratification: that is to say, the positions of highest status, power and prestige are occupied on the whole by men, whereas women tend to be found in positions of low status, low pay and low security.[3] Thus to find women practitioners, we must be prepared to look away from the elite forms of practice, those with the highest public profile and which spring most readily to mind.

Our conception of the history of women as medical practitioners has been dramatically enlarged over the last two decades. Twenty years ago, the subject was largely confined to the great pioneer women doctors of the later nineteenth century – Elizabeth Blackwell, Elizabeth Garrett Anderson and Sophia Jex-Blake – and their heroic struggles to obtain medical training and qualifications against the prejudice, bigotry and deliberate obstruction of universities and medical corporations.[4] A larger picture began to be developed during the 1970s, mainly as a result of the Women's Movement. In the practice of 'consciousness-raising', women discovered shared bodies of experience which had been ignored or trivialized by men, and this gave rise to a programme of women's history for the recovery of past women's experience which had been ignored or trivialized by male historians.[5] It was discovered (or rediscovered) that women had practised as surgeons, apothecaries and even as physicians at all times from the Middle Ages to the eighteenth century, although their activities seemed to have been gradually curtailed.[6] There was study of midwives, who from the start of the eighteenth century faced increasing competition from medical 'man-midwives' claiming exclusive authority for the management of childbirth.[7] There was new awareness of the importance of the healing done by women within their family circle or by those known as 'handy women' or 'old wives' and other irregular practitioners; for it was such people, and not the medical profession or any recognizable antecedent of it, who treated most of the population before the urban industrial society of the nineteenth century.[8] During the 1980s, the history of nursing was raised to prominence by women's historians, in conjunction with social historians interested in the health-care services, and with present-day nurses interested in the origins of their current professional problems.[9] It was becoming clear that the history of women as medical practitioners had to encompass irregular or less prestigious practice, such as that of old wives, midwives and nurses.

But, at the same time, the Women's Movement was enabling us to interpret the low status accorded to women's medical practice in a different way. Twenty years ago, medicine itself was seen as neutral and gender-free; the only question to ask about the history of female practitioners was to what extent women had been able to practise medicine at different times in the past, and the only conceivable explanation for gender stratification was sexual discrimination. The development of feminist analysis, however, led to the challenging of those very values by which the hierarchies of knowledge and occupation were established. From the perspective of the Women's Health Movement, for example, suspicious of modern medicine as an instrument of men's power over women and aiming to enable women to re-establish control over their own bodies, midwives and old wives of the past were figures of inspiration; their persecution and vilification as ignorant and superstitious (opprobrium still attaches to the phrase 'old wives' tales') was a sign not of their incompetence but of their oppression.[10] Less spectacularly, but no less radically, research on nineteenth-century nursing revealed that the subordinate role of the nurse was not inevitable; it was doctors who wanted the new nurses to have only such training as to be more efficient assistants for themselves, whereas matrons and nurses campaigned for independence, autonomy and professional status, and a general education to match.[11] Research on the women doctors of the pioneering generation showed that most of them believed that they had a distinctive contribution to make to medicine, a particular role by virtue of their 'natural' female gifts as nurturers and healers. Their practice in areas of low prestige and low income – in out-patient dispensaries for the inner-city poor, in poor-law infirmaries and other public-sector institutions, in India treating Muslim women – was to some extent forced on them, but it was also the case that they placed a high value on this kind of practice; it was only according to conventional values that the treatment of the poor and disadvantaged carried less status than exclusive private practice or hospital consultancy.[12] As in all this research, the priority on recovering women's experience and appreciating their values led to the reinterpretation of whole areas of women's practice and knowledge; they were now seen as not so much *inferior* as *female*.

This chapter is an attempt to reinterpret women's work in sanitary reform along these lines. Sanitary reform is an aspect of women's medical practice which has attracted very little attention,[13] partly because of the low status conventionally accorded to it as preventative medicine, though in its educative aspect it was strongly associated with the prevailing forms of femininity. Thus if we are looking for characteristically female forms of healing practice and female forms of knowledge, then sanitary reform must be a good candidate. This chapter will therefore examine the nature of women's work in sanitary reform, the reasons why this practice arose, and the extent to which it was not merely a women's practice but a female practice. There will be one major omission: the work of Florence Nightingale. Although she is now generally associated with nursing, in fact her only lasting interest in nurses was for the role they might play as agents of sanitation.[14] Most of her career was spent in the design of healthier hospitals,[15] in the reform of the army medical service, and in the reform of the sanitary administration in India.[16] But while she was by far the most successful woman sanitary reformer in Victorian Britain, she was quite exceptional in working, to say nothing of becoming a national authority, in the male areas of engineering and government administration. Her sanitary work really demands an article to itself.

This chapter will therefore deal only with the more usual sanitary work of middle- and upper-class women. The first section will explore what kind of activity it was, and in particular its relation to philanthropy. The second will examine the kind of knowledge which the women sanitary reformers were trying to impart, as revealed by the records and publications of the Ladies' Sanitary Association. The conclusion will outline the reasons for the demise of women's sanitary reform, and discuss some of the implications of this for our view of modern medicine.

SANITARY REFORM AS PHILANTHROPY

The Victorian movement for sanitary reform had two essential character-istics.[17] First, it was a movement not of doctors but of lay people. Second, it was concerned with the health not of people in general but the working classes specifically. It began to flourish during the 1840s, at a time when middle- and upper-class people were increasingly alarmed by the rise of Chartism and other forms of working-class activism, and were urgently seeking to identify and redress possible causes of discontent amongst the lower orders. The 'Condition of England' question, as it was often called in the periodical press, continued to remain one of the leading social and political issues of the time, even after the collapse of the Chartist movement, and sanitary reform was considered a major part of it. That this was so was largely the work of one political group, centred around the Benthamite administrator Edwin Chadwick, which since the early 1830s had been using its access to government apparatus to publicize the ill health and filthy living conditions of the working classes, through such publications as the 1842 *Report on the Sanitary Conditions of the Labouring Population of Great Britain*. The Benthamites had their own reasons for being concerned about the ill health of the working classes – it caused loss of labour, and hence loss of national wealth and a drain on the Poor Rate – but once they had given currency to the concepts of 'public health' and 'sanitary condition', it became essential for anyone concerned about the working classes, for whatever complex mixture of reasons, to give attention to sanitary reform.[18] Thus when politicians, administrators, philanthropists, social theorists and 'all those interested in social improvement' came together in 1857 to found the National Association for the Promotion of Social Science, amongst the sections devoted to such standard Benthamite or Liberal subjects as Law Amendment, Education, Prevention and Repression of Crime, and Social Economy, was a section devoted to Public Health.[19] In that section were discussed such matters as the working of the 1848 Public Health Act, the role of municipal government in improving urban sanitation, the engineering and chemical technology of sewage disposal and treatment, and water pollution. The Ladies' Sanitary Association was affiliated to the NAPSS, and to some extent its members shared the Public Health section's interest in legislation, administration and engineering. Fundamentally, however, it represented a different approach to sanitary reform, based on the belief 'that in a majority of cases the principal cause of a low physical condition is ignorance of the laws of health.' The aim of the Ladies' Sanitary Association was 'to propagate this important branch of knowledge'; in other words, its role was one of public education.[20]

Middle- and upper-class women moved into sanitary reform because they were already active in philanthropy, and the philanthropic tradition was crucial to their conception of their aims and methods. For example, as Mary Anne Baines explained to the NAPSS, although 'to improve the physical condition of the poor' was the Ladies' Sanitary Association's 'primary and more direct object',

> another benefit is likely to result, not less important in its character – an improvement in the moral and social condition of the working classes, brought about through the natural relation that exists between the physical state and the moral condition, . . . but much more is hoped in this case from the sympathy and interest of educated women being thus elicited towards their poorer neighbours.[21]

This emphasis on personal contact was confirmed by the subsequent development of the movement. Although the initial programme of the Ladies' Sanitary Association included ambitious plans for teacher-training centres and loan libraries, the most successful activities were the standard ones in philanthropy: publishing tracts, organizing lectures, and above all visiting. This pattern of activity had been developed mainly by evangelical and Nonconformist women, for whom commitment to 'serious' religion – more than mere ritual and doctrine – meant that faith was to be shown through works.[22] From their position of middle-class respectability, they were appalled by the physical, the moral and the religious condition of the working-class poor, and through philanthropy they aimed to improve all three simultaneously. Whichever might originally be uppermost in a philanthropic visitor's mind, before long she generally found herself giving roughly equal weight to material aid (such as food, clothing or medical treatment), moral advice (on good manners, temperance and thrift) and religious instruction. The Ladies' Sanitary Association, in Mary Anne Baines's account, was explicit in combining the physical and the moral in its aims, and at a local level the movement encompassed the religious also, for the women most ready to convey sanitary knowledge to the poor were the ones who were already engaged in conveying the Gospel. Many local branches of the Ladies' Sanitary

Association worked in co-operation with the network of working-class Bible women organized by Ellen Ranyard, so that the Bible women became sanitary as well as Christian missionaries.[23] This pursuit of the physical, moral and religious 'improvement' of the poor through personal contact and instructional literature was absolutely characteristic of women's philanthropy.

We are only just beginning to understand the scale and importance of women's philanthropic activity during the nineteenth century. We have tended to regard it as one of the amusements of well-meaning but ignorant ladies with time on their hands, very much on a par with embroidery, paying calls, reading aloud and the other apparently trivial occupations of middle- and upper-class women. In one respect, this view was quite accurate: women's philanthropy was seen as part of women's domestic role. According to the dominant ideology of 'separate spheres', man's place was in the harsh competitive sphere of business, and woman's place was in the peaceful, tender sphere of home and family, but it could be legitimate for this role of caring and nurturing to be extended to others outside the home in the form of philanthropy or, as Marina Benjamin discusses, pedagogy. In fact, during the first half of the nineteenth century, philanthropy was virtually the only acceptable way for middle- or upper-class women to be active outside the home, and what we are beginning to appreciate is that it provided them with new and positive opportunities, which many of them therefore seized with both hands.[24] Even in district visiting or Sunday-school teaching, a woman could become a social individual and a dispenser of patronage in her own right. Campaigning or fund-raising for a local or national cause, notably the abolition of slavery, brought further opportunities for administration and management, as did service as a 'lady visitor' to factories or public institutions, pressing for humane conditions following the lead of Elizabeth Fry (prisons), Mary Carpenter (reformatories for juvenile offenders) and Louisa Twining (workhouses). Otherwise excluded from formal intellectual education, politics and paid work, middle- and upper-class women could, in philanthropy, develop skills, exercise power, and do work which they could feel to be useful.[25]

It was for this reason that Victorian feminists gave philanthropy a leading place in their campaigns to promote 'women's work'. Many of them, such as Barbara Leigh Smith (later Barbara Bodichon), came from evangelical or dissenting families in which 'WORK – not drudgery, but WORK –' labour in the service of some higher cause, was regarded as

'the great beautifier': an expression of faith, ennobling and dignifying.[26]
As this passage from her *Women and Work* illustrates, their demands for
women's freedom from domestic idleness were based on this view of
work:

> One duty in this world is to try and make it what God intends it shall
> become: we are His tools. By working for the salvation of this world, we
> may chance to achieve our own in another, but never by any other means. . . .
> To do God's work in the world is the duty of all, rich and poor, of all
> nations, of both sexes. . . . Women must, as children of God, be trained to
> do some work in the world.[27]

This was a concept of work which went far beyond the economic.
'Women and Work' was first published in the *Waverley Journal*, which
Barbara Leigh Smith and her close friend Bessie Rayner Parkes, also
from a Unitarian family, took over in 1857 with the aim of turning it
into a 'Working Woman's Journal'. The term was intended to include 'all
women who are actively engaged in any labours of brain or hand' – that is
to say, 'the many . . . labourers in the broad field of philanthropy' as
much as teachers, artists and manual workers.[28] The more famous *English
Woman's Journal*, which they started the next year when it proved
impossible to buy the *Waverley* outright, is usually remembered for its
pioneering feminist analysis of women's economic, legal and educational
disadvantage in British society, and its role in gathering together women
who campaigned in a variety of women's causes during the 1860s and
1870s. In fact, the *English Woman's Journal* retained the *Waverley's*
central focus on women's work, including philanthropy, and the
possibilities for expanding the field. Women's philanthropic work was
thus seen as an area of great excitement, offering a great sense of
purpose.[29]

The sense of excitement was heightened rather than diminished by the
principle of 'separate spheres', that women and men had different
qualities and abilities, and were ordained for different spheres of
influence. This principle with respect to philanthropy was expressed most
influentially by Anna Jameson, a highly successful and respected writer on
artistic and religious subjects, who was something of a mentor for Bessie
Rayner Parkes and Barbara Leigh Bodichon.[30] By the 1850s, Jameson
was turning her attention to women's work in philanthropy, and in a
lecture of 1856 she warned that 'some of our noblest educational and

charitable institutions' had 'sundered what in God's creation never can be sundered without pain and mischief, the masculine and the feminine influences,' so losing 'the true balance between the element of power and the element of love.'

> It appears to me that the domestic affections and the domestic duties – what I have called the 'communion of love and the communion of labour' – must be taken as the basis of all the more complicate social relations, and that the family sympathies must be carried out and developed in all the forms and duties of social existence, before we can have a prosperous, healthy, happy, and truly Christian community. . . . All our endowments for social good, whatever their especial purpose or denomination – educational, sanitary, charitable, penal – will prosper and fulfil their objects in so far as we carry out this principle of combining in due proportion the masculine and the feminine element, and will fail or become perverted into some form of evil in so far as we neglect or ignore it.[31]

Such sentiments were encouraging to women, for what they implied was that here was work which *only* women could do. Women's qualities and abilities were thus given an extremely high valuation: that of being absolutely necessary if failure or evil was to be avoided. The women's moral crusades of the 1870s – the campaign against the Contagious Diseases Acts led by Josephine Butler, and the campaign against vivisection led by Frances Power Cobbe, which Lynda Birke discusses below – demonstrate spectacularly how a belief in essential female qualities could provide a great sense of power which enabled women to face and overcome obstructions and difficulties.[32] Something of the same power was involved in women's work in philanthropy, as set out by Anna Jameson.

To locate the women's sanitary reform movement within the realms of women's philanthropy is therefore to locate it within a powerful and exciting female tradition. It was a new accession to the expanding sphere of women's work, and as such received considerable coverage in the *English Woman's Journal*. Indeed, during its early years, the Ladies' Sanitary Association shared the *Journal's* offices, and the *Journal* was officially constituted the Association's organ, thus allowing sanitary publications, appearing there first, to be later used by the Association without the expense of re-setting the type.[33] One of the first tracts to be published in this way was Susan Rugeley Powers's 'The Details of Women's Work in Sanitary Reform', which carried at its head Anna

Jameson's affirmation of the necessity for the feminine element in all social projects. The same quotation was placed on the cover of each of the Ladies' Sanitary Association's annual reports, thus serving as a reminder of the importance of the work which was waiting for women to do. For, as Bessie Rayner Parkes wrote:

> The best framed Acts of Parliament most efficiently carried out, will only result in partial reforms, until the habits of the people, engendered amidst bad conditions, and rendered careless by hopelessness, be also changed. . . . We want just that minute domestic instruction in matters sanitary for which Boards of Health may pave the way, but which they can never complete in detail. We want the action of *women* in every parish; we want the clergyman's wife and the doctor's daughter to know the laws of health, and to enforce them in the perpetual intercourse which we hope and believe they maintain with their poorer neighbours. The squire's lady, and the peeress whose husband owns half the county, the district visitor who cares for the soul, and the parish nurse who attends upon the sick – if all these women could be made to work with a will . . . what a difference might be wrought in the average mortality of England.[34]

We shall now look in more detail at the methods of women sanitary reformers and the nature of the knowledge they were trying to impart – in particular, the meaning of that crucial but cryptic phrase, 'the laws of health'.

SANITARY REFORM AS GOOD HOUSEKEEPING

The programme of sanitary reform, as laid out by Edwin Chadwick and his radical physician colleagues, drew on the classical medical tradition in which changes in the air (one of the Galenic 'non-naturals') was a principal cause of mass disease.[35] They saw the typhus, yellow fever, cholera and other fevers endemic and epidemic amongst the working classes as the result of the poisoning of the air by 'effluvia' or 'miasms' rising from decomposing organic refuse and stagnant water. This has been commonly described as a 'miasma' theory of disease, as opposed to a 'contagion' theory in which disease is transmitted from person to person, but this is now generally regarded as a false dichotomy; before the advent of the germ theory of disease, there was general agreement that many if not all diseases could *both* be communicated by direct contact *and* contracted from atmospheric influence, such as miasms arising from the excreta of a diseased person.[36] The early sanitary reformers' reasons for focusing exclusively on atmospheric causes were twofold. First, these causes were extremely evident; the pervasive overpowering stench of inner-city slums was taken to be a clear sign of the presence of poisonous effluvia in the air. Second, the main sources of these effluvia – heaps of excrement in the streets, overflowing drains, badly designed and over-used privies – were extremely remediable. Action could be taken simply on the basis of the observed correlation between filth and disease, and Chadwick and his colleagues used official mortality statistics and the testimony of official inspectors to establish this correlation and make it impregnable. Although some doctors and chemists attempted to investigate the chemistry of the pathogenic changes in the air, generally using an analogy with the process of fermentation (the fevers in question were often referred to as 'zymotic' or yeast-like diseases), the early sanitary reformers mostly steered clear of detailed theories of disease causation, and frequently regarded discussion of such theories as a diversion from the task which needed to be done, namely, to prevent the accumulation of impure air.[37] From this perspective, it followed that the main aim of sanitary legislation and engineering – the men's side of sanitary reform – would be with the removal of 'nuisances' (such as accumulations of excrement), the building of better drains and sewers, the redesign of houses to allow for good ventilation, and the provision of clean water – not so much for drinking as for washing.

Women sanitary reformers by contrast addressed a rather broader range of causes of ill health. They shared the general concern with fresh air and cleanliness; hence the Ladies' Sanitary Association published tracts on subjects such as ventilation (*The Cheap Doctor: A Word About Fresh Air*, *The Mischief of Bad Air*, *The Worth of Fresh Air*), washing (*The Power of Soap and Water*, *The Use of Pure Water*), and general hygiene (*Black Hole in our Bed Rooms*, *Village Work*, *Whose Fault Is It? Sanitary Defects*, *How do People Hasten Death? The Secret of a Healthy Home*, *Rules for the Sanitary Arrangements of Pig Styes*).[38] But the Ladies' Sanitary Association also published tracts on diet (*The Value of Good Food*, *The Influence of Wholesome Drink*, *Cookery for the Million*), and several of its branches ran cookery classes and campaigned for the teaching of cooking in elementary (that is, working-class) schools. 'The amount of suffering from indigestion, caused by improperly prepared food, from the perpetual use of tea, as well as of alcoholic liquors, is hardly to be credited,' wrote a correspondent from the Ladies' Branch of the Manchester and Salford Sanitary Association.

> The reason seems to be that girls go from school to the mills without ever having opportunity for learning practically, cooking, or anything that may teach them how to make their homes happy and healthy. The result is that they live chiefly on tea and bread and butter, which save trouble, and on extra occasions they purchase ready-cooked food of an expensive and indigestible nature.[39]

In a similar vein, the Ladies' Sanitary Association during the 1880s was involved in a campaign for wholemeal bread.[40] Women sanitary reformers were also concerned with clothing, and the health risks of 'tight corsets, tight sleeves, weighty skirts, high-heeled boots, tight stiff collars, ill-adjusted braces, pointed toes' – all laid out in the graphically-titled tracts *On Dress: Its Fetters, Frivolities, and Follies* and *Wasps* [that is, wasp-waists] *Have Stings; or, Beware of Tight Lacing*. The Ladies' Sanitary Association published patterns of model garments to promote the principles of hygenic clothing ('allow the vital organs unimpeded action, suspend the clothing from the shoulders, reduce the weight as much as possible, and preserve a uniform temperature over the body').[41] Finally, and most importantly, women sanitary reformers were concerned with maternity and child care. From the start, the Ladies' Sanitary Association devoted much effort to this area, its tracts on maternity including *Evils Resulting from Rising too Early after Childbirth* and *The Health of Mothers*; while on child care there was *Washing the Children, How to Manage a Baby, Hand Feeding, Little Mary's Illness, The Sick Child's Cry*, and *How to Rear Healthy Children*; and for publicizing the enormous scale of preventable infant mortality there was *The Massacre of the Innocents*. There was also a campaign to discourage working-class women from hiring themselves out as wet-nurses, which frequently led to their own babies being undernourished. It would be an exaggeration to say that these matters were of no interest to men sanitary reformers, but they were not, in general, a major part of their concerns. Women sanitary reformers, however, had as their aim the reform of working-class women's practice across the whole field of housekeeping. As women involved with housework, cooking, clothing and child care themselves, they could not, like men, ignore these things or regard them as unimportant.

Women sanitary reformers were well aware that if they were to get their important information across to working-class women, the presentation would have to be popular, interesting and memorable. The tracts of the Ladies' Sanitary Association show a variety of devices being adopted to this end. Some of them were written in verse – short pithy, rhythmic lines, calculated to capture the attention and stick in the memory. One set of verses, *Hints on Your Health*, was supplied mounted on cardboard with eyelet holes, ready for hanging on a wall. Other tracts were cast into the form of stories, though in general the plots were pretty thin, the characters acting as talking heads. Even here, though, the setting in which the dialogue took place was chosen with some care, to build confidence. For example, in *The Bride's New Home*, the recently-married Mary Gray is visited by her neighbour, the older Mrs Foster, who on her own marriage had received good advice from Mary's mother (now dead), thus establishing a context of women sharing tried and trusted information with each other ('Your mother said to me . . . and rely upon it she was right'). The information itself was wherever possible made more vivid by graphic and arresting imagery. For example, in the same tract, the lesson about the importance of proper diet is taught through a fantasy about what your stomach would tell you, if only it could speak.

> If you send into me such ill-cooked substances that my acid cannot dissolve them, or if you pour into me such quantities of porter or ardent spirits as to destroy or burn up the fine stuff of which my coats are made, how can I do my work properly? . . . And yet no sooner do you feel the inconvenience of your own acts, than you find fault with me; and forsooth, it is the stomach that is out of order. That may be true; but please let me ask you the question – Whose fault is it that it is so?[42]

Later in the same tract, the lesson about the importance of fresh air is taught through a cautionary tale. Mary's husband goes to help his brother, a gardener for the Squire, dress the Hall with flowers for a big party. The morning after, they come to remove the flowers, and find them completely wilted. 'What has done it?' asks Mary's husband.

> 'It is the bad air that has done it,' answered his brother; 'it is always the same. I dread to see my plants come to these grand parties, for mistress invites such a number of people that they crowd the rooms till every atom of fresh air is breathed up, and then of course as they must breathe something, they breathe bad air and worse air out, till it is little better than poison. . . . I

heard the other day of a large party given in a small room. They danced the whole night, keeping doors and windows tight shut. Well, what was the consequence? Seven people were down almost immediately of the typhus fever, and then folks said, 'Dear me, how terribly typhus fever is about'! as if it would have been about if they had not brought it themselves.[43]

The object of such stories was not just to shock, or to tell the working classes that they brought ill health on themselves, but to motivate them to adopt more healthy habits. Thus all the tracts of the Ladies' Sanitary Association included extensive discussion of the practical details of daily life. The tracts on infant management included detailed advice on such subjects as bathing (the right water temperature, the parts not to miss, how to dry off); feeding (suckling, bottle feeding, weaning, not giving solid food too soon); and sleeping (especially on not giving sleeping draughts).[44] Tracts on diet included tips on shopping and cooking, with complete recipes. A striking example of the practical detail of these tracts is to be found at the end of *The Bride's New Home* where we are taken into the Grays' bedroom – a model for the reader to imitate.

> In one corner stood an iron bedstead – no dirty hangings – no close head-piece, but simply a painted iron framework, upon which was laid a mattress, and light-looking bed clothing. There was a small chest of drawers, and a row of pegs along the wall, on which hung Will's coat, and Mary's gown and cloak. There was a small piece of carpet by the side of the bed; the rest of the floor had no covering, and the boards looked as clean as soap and water could make them. The window was open, and the fire-place, though small, was open likewise; so that, as Mary said, this room at any rate was enjoying the luxury of a thorough draught.

Subsequent dialogue explains and adds to the details. The space under the bed is kept clear and clean so that fleas do not breed; Will and Mary have rejected a feather bed, as weakening to the body; the hanging pegs allow their clothes (warm and light) to air.[45] From what we know of sanitary visitors, they too concentrated on practical details; indeed, it was the only way to begin a visit. Talking about the children was found to be a good way to break the ice, from which the visitor might move on to, 'If I were you, Mrs.—— I would open the windows a little. Shall I do it for you?'[46] Discussion of general sanitary principles was more effective if the recipient could already feel for herself some of the benefits of fresh air and cleanliness.

Even at the most abstract and intellectual level, practicalities were interwoven with physiological theory. This is illustrated by the programme of a series of lectures given by a Miss Armstrong in Nottingham in 1881, organized by the local branch of the Ladies' Sanitary Association:

I. *Bone and Muscle.*– Blessing of Good Health. Why so many have Bad Health. Frame-work of the Human Body. Its Division into Three Parts:– Skull, Trunk, Limbs. The Joints. Necessity of handling Children gently. What Bone is made of. Causes of Deformity. Movement. Clothing.

Flesh consists of Muscle. Its two uses. What it is made of. Exercise.

II. *Breathing.*– Air the first necessary of Life. What it is made of. Difference between the Air we breathe in and the Air we breathe out. Resemblance between Breathing and Burning. Description of the Lungs. Consumption. Warm Clothing.

III. *The Blood and its work.*– Good Blood and bad Blood. What happens to the Blood in the Lungs. The Heart and its action. How the Blood is carried all over the Body, and what it does there. Tight Clothing. Out-door Exercise.

IV. *Why do we eat?*– What is Food? Its uses in the Body. The Alimentary Canal. The Liver. Digestion and Assimilation. Indigestion. Eating slowly. Regular meals.

V. *What is the use of cooking?*– Four Classes of Food. Milk. Some other principal Foods. Five things that a real knowledge of Cookery teaches. Economical Dinners.

VI. *Water.*– Its Existence in Nature. Water as a Drink and as a Cleanser. Rain. Danger of Poison in Water. How to be avoided. Hard and Soft Water. Characteristics of a good Drinking Water.

VII. *Other drinks.*– Why do we drink? The use of Water in the Body. Nourishing Drink: Milk, Cocoa.

Stimulating Drinks: Tea and Coffee.

Intoxicating Drinks: Beer, Wine, Spirits.

Fruit syrups, &c.

VIII. *The work of the lungs, the skin, and the kidneys.* – Why these are called the Excretory Organs. Second use of the Lungs.

Two uses of the Skin. Its Formation. Perspiration. Why Cleanliness is necessary. The Hair and the Nails.

The Kidneys and their work. Resemblance between work of Lungs, of Skin, and of Kidneys.

IX. *The brain and the nerves.*– The Brain the Ruler of the Body. The Spinal Cord. Connection of Brain and Nerves. Nerves the Messengers of the Body. The Organs of Sense. Some of the Causes of Nervous Illnesses. Paralysis. Nervousness of any Invalid. Necessity of treating Children gently. Froebel. Change. Exercise. Rest. Sleep.

X. *Healthy homes.*– Drainage: The Poison in Drains. Some of the Diseases brought on by Bad Drainage. Consequent necessity of understanding the Drainage Arrangement of our Houses. Some of the common Faults in these Arrangements.

Ventilation. Air the great Purifier. How to get Fresh Air without Draughts.

Dryness: Can the Damp be kept out?

Warming and Lighting: The economical use of Coal gas.

XI. *Communicable diseases.*– What is Infection? Excessive fear of it. Carelessness about it. Children's Infectious Diseases. Germs in Air and Water. Small-pox. Typhoid. Difference between Typhoid Germs and others. Disinfectants and Antiseptics. Management of Room and House where Infectious Patient is ill.

XII. *Further management of sick room and of all bed rooms.*– (For women only.)

XIII. *Personal health.*– (For women only.)[47]

Even from the synopsis, it is clear that each lecture moved constantly back and forth between the theoretical and the practical: the subject of muscle led to the subject of exercise, blood to the evils of tight clothing, the physiology of digestion to proper eating, the skin to cleanliness. For women sanitary reformers, there was no point in discussing the general principles of sanitary reform unless they were related to practical details.

The practicalities under consideration were not just physical. Discussions of diet, for example, emphasized the connections between digestion and mental states: how bad temper and anxiety produces indigestion, and healthy eating produces good temper. This was not just psychosomatic medicine; there was a very direct moral point being made, as we can see from *The Bride's New Home*:

> If a woman would only answer the question honestly, she might trace many an angry word from her husband to a greasy, frizzled-up steak, or a meal of bread and cheese, helped down with ale or porter, or perhaps with gin, or rum, or whisky; whilst if the wife would but have given herself the trouble to prepare a good, hot, nourishing dinner or supper for her husband, his temper would not have been irritated, his stomach would not have been put out of order, and the temptation to intemperance would at once have been removed.[48]

Proper diet was thus necessary for the production of not only sound physical health, but also respectable family life. Mary Gray, the model character, has naturally no difficulty here, and at the end of the tract we find her preparing an excellent supper for her husband. On the last page, he comes home from work, kisses her, and after supper, while she clears away, he says, 'Bless you, darling; I really do not think any man ever had such a good wife as I have!' This utterance was the punchline of the whole tract: fundamentally it was about how to be a 'good wife'. It was one of a series, published early in the career of the Ladies' Sanitary Association, with the general title *Wedded Life: Its Duties, Cares and Pleasures*. Other tracts in the series dealt with the care of infants and the bringing up of children, encouraging the use of moral pressure rather than physical coercion, and advocating the building of a relationship of trust and authority. The final tract, with the model, well-behaved children now going to school (and winning prizes), concluded with an explicitly Christian peroration: 'Men and women, do you want such a home as the Grays had – such children as the Grays had? Follow their example. Serve God heartily – serve God faithfully; bring up your children in His fear, and rest assured you will not miss your reward.'[49]

This easy combination of physical, moral and religious instruction was not peculiar to these particular publications, or to the home visitors who distributed Bibles as well as sanitary tracts, or the minister's wives who held home meetings which began with sewing, went on with sanitary lectures, and closed with prayer;[50] as we saw in the last section, this combination was characteristic of women's philanthropic work. Further-more, it was built into the concept of 'the laws of health' (or sometimes 'the laws of life') which was basic to women's work in sanitary reform. In trying to understand this concept, it is important to realize that the term 'law' has for us two distinct meanings; a scientific law, by which one event follows another in a determinate way, is quite different from a law promulgated by a monarch, a parliament, or other authority. Victorian sanitary reformers, however, invoked both these meanings simultaneously. Thus, for example, a Miss Wyndham explained the purpose of women's work in sanitary reform as follows:

What we ladies claim to do, is . . . to explain to the women [of] poor neighbourhoods . . . how much they suffer from ills the good God has never sent them, but which are the outcome of their own ignorance and idleness, or want of personal care or care of their children. . . . I try to impress upon them that our Heavenly Father has created this world, as well as the next, and that He governs by great physical laws, which if we study and do not break, we cannot but acknowledge to be for our good, and whose ultimate benevolence we must recognise, although they punish ignorance and neglect so severely.[51]

The 'laws of health' were *both* the laws of physiology and chemistry, by which ill health was the natural consequence of an unhygenic lifestyle, *and* the commandments of a divine lawgiver, commandments which it was possible to disobey.

This was not just a confusion resulting from ignorance or sloppy thinking. As is hinted in Miss Wyndham's statement, it was the product of a deliberately-adopted theological position: that evil (such as sickness or death) was a natural consequence of generally good physical laws. The importance of this position was not so much that it offered a solution to the problem of evil, but that it implied a certain attitude towards misfortune: that one should not accept suffering fatalistically, but should take action to discover and circumvent its cause, and so improve one's own condition. This theological position and its implications were clearly stated in one of the Ladies' Sanitary Association's verse tracts entitled *Village Work*. The tract describes a village, the rural idyll of which is marred by the ill health of its inhabitants. Their energetic new vicar recognizes the problem, and determines to do something about it:

> For he had looked about their homes,
>   And traced full well the cause;
> 'Twas plain enough, they'd never heard
>   Of sanitary laws.

He has great difficulty in convincing the villagers that they can and should do anything about it.

> For none believed him when he spoke
>   Of cause and of effect;
> It seemed so strange to hear him say
>   'Twas from their own neglect.
> Some had a notion it was *wrong*
>   To trace out cause of ill;
> Submission was the Christian's part –
>   Submission to God's will.

One man in particular argues with the vicar, even though he has lost many children through fever and is in ill health himself.

'I don't know what comes over me,
  I lose my strength of limb;
But there, it's all God's will we know,
  And we must trust to Him,
And bide our time; if He thinks best,
  We *must* be weak and ill;
Do what we may, 'tis all the same,
  We cannot change His will.'

The vicar argues with him.

'We cannot change God's will, you say,'
  The Vicar answered: 'True;
Then let us strive to *learn* His will
  In everything we do.
In common cares of common life
  He teaches day by day;
By common laws He rules the world –
  Those laws we must obey. . . .
For instance,' said he, 'do you think
  When men drink night and day,
And lose their health and manly strength,
  They have a right to say,
"It is God's will," whilst they go on
  Still drinking just the same?
Is it God's will that men should live
  In drunkenness and shame?
Not so; it is the penalty
  Men pay for drunkenness;
And as that brings its punishment
  (As well as all excess),
So will neglect of cleanliness
  In household matters bring
Its lesson, or its punishment,
  By death, or suffering.'

The villagers are won over, and under the vicar's guidance, and with the help of the squire, the village is cleaned and drained, and it becomes a healthy and a happy place.[52] The moral of the story is that humans can and should improve their own condition through knowledge of the physical laws which God has ordained, and above all (as the title of the tract reminds) through work.

Thus when middle- and upper-class women sanitary reformers described their aim as spreading knowledge of 'the laws of health', the knowledge which they had in mind was neither neutral nor abstract. Their programme was based on the view that God did not visit punishment or reward on people by some kind of miraculous intervention but, like an ideal *laissez-faire* government, stood back and let natural laws have free operation. Like the laws of the market place, the laws of nature were believed to be ultimately beneficial and productive of general happiness, but it was up to each individual to be aware of them and take action accordingly to help themselves. It was the object of women sanitary reformers to equip the working classes for this kind of self-help by educating them in the laws relating to health – physical, moral and religious. The full name of the Ladies' National Association for the Diffusion of Sanitary Knowledge recalled two earlier foundations: the Society for the Diffusion of Useful Knowledge and the Society for Promoting Christian Knowlege; and this connection was probably deliberate. Sanitary knowledge, like the scientific and literary knowledge which was the business of the SDUK, was intended to provide the means for a practical improvement in the lives of working-class people; and like the scriptural knowledge of the SPCK, it was intended to provide the means of salvation – reckoning the achievement of respectable independence as a form of social salvation. Like the other two societies, then, the Ladies' Sanitary Association aimed to supply the working classes with useful, saving knowledge, which would enable them to help themselves and improve their own condition.

THE END OF SANITARY REFORM

Women's work in sanitary reform, as exemplified by the Ladies' Sanitary Association and its affiliated local organizations, did not continue much beyond the mid-Victorian period. As we have seen, the practice had its roots in women's philanthropic work aimed at encouraging working-class self-help; but by the 1890s the home visiting previously done by unpaid philanthropists was being done by salaried municipal employees. Again, the knowledge-base of women's sanitary reform depended on the principle that good health was a product of good housekeeping; but from the 1870s onwards the rich theories of disease causation, which had roles for air, diet, clothing and emotional state, were steadily displaced by a reductionist germ theory. Both of these changes have made it harder to recover the nature of women's practice in sanitary reform and to take it seriously. The assimilation of women sanitary reformers into the ranks of municipal officials brought them into direct professional conflict with men sanitary inspectors, who sought to confine women to the newly-created rank of 'health visitor' and to establish an inferior status for this role. Their deeply interested and transparently gendered representation of women health visitors as 'looking at things out of perspective', 'magnifying a feeding bottle' and being paid to go 'gossiping from door to door' has tended to persist.[53] Similarly, with the increasing adoption of germ theory, the importance and even the possibility of women's sanitary work was severely diminished. Disease became identified with the presence of a microbe which could only be detected in a laboratory by a highly-skilled experimenter; it was no longer possible for lay people to detect the presence of pathogens through their God-given sense of smell,[54] or to educate themselves to expertise in sanitation from home, as Florence Nightingale had done. The moral and religious component of sanitary theory – so essential to the mid-Victorians – was eliminated, and all sanitary measures which could not be seen as preventing the transmission of germs were relegated to secondary importance.[55]

The rise of germ theory in the last few decades of the nineteenth century was just one part of a more general trend: the increasing dependence of medicine upon the laboratory. The new world of laboratory medicine was very much a man's world, not only in the sense that it was (and is) occupied predominantly by men, but more profoundly in the sense that men established its values and priorities in accordance with the values and priorities claimed by men in this culture. In regarding the laboratory as the prime site of knowledge production, men valued abstract theoretical knowledge over concrete and practical. By making laboratory tests a major form of diagnosis, men gave preference to the reduction of complex patterns of disease to single elements more easy to dominate and control. Through the widespread adoption of laboratory-derived therapeutic techniques, men's favoured model of treatment become expert intervention in the manner of an experimenter. It is precisely these masculine values and priorities of modern medicine of which we are now becoming critical.[56] For this reason, I have found it both interesting and exciting to find out about a form of medicine which was developed by women in accordance with their own values and priorities. Women's sanitary reform was a distinctive practice of the mid-Victorian period, and it would not be possible to resurrect it today, even if we wanted to. (Most of us are not as sanguine as the Victorians about what self-help can achieve.) Yet the work of women sanitary reformers can still serve as a reminder that the practices which we inherit are not fixed and unchanging; that we can indeed appreciate lay preventative medicine, knowledge that is personal, practical, popular and empowering, values that are acknowledged and not seen as weaknesses. It can be a reminder of possibilities.

ACKNOWLEDGEMENTS

My thanks to Troy Cooper, who taught me feminism; Andrew Cunningham, who taught me history; and Johanna Geyer-Kordesch, who suggested the subject of this chapter.

NOTES

1 S[usan] R[ugeley] P[owers], 'The details of woman's work in sanitary reform', *English Woman's Journal*, 3 (1859), pp. 217–18.

2 I am not counting the Epidemiological Society, founded in 1850, which was not precisely sanitary in its scope. The Sanitary Institute of Great Britain, for administrators and engineers, was not founded until 1876.

3 Mary Ann Elston, 'Women in the medical profession: whose problem?' in Margaret Stacey, Margaret Reid, Christian Heath and Robert Dingwall (eds), *Health and the Division of Labour* (Croom Helm, London, 1977), pp. 115–40. For the parallel situation in the USA, see Carol Lopate, *Women in Medicine* (Johns Hopkins Press, Baltimore, 1968).

4 Sophia Jex-Blake, *Medical Women: A Thesis and a History* (Oliphant, Anderson and Ferrier, Edinburgh, 1886); Elizabeth Blackwell, *Pioneer Work in Opening the Medical Profession to Women* (Longmans, London, 1895); Margaret Todd, *The Life of Sophia Jex-Blake* (Macmillan, London, 1918); Ray Strachey, *The Cause: A Short History of the Women's Movement in Great Britain* (Virago, London, 1978; original edition, 1928), chs 9 and 13; Louisa Garrett Anderson, *Elizabeth Garrett Anderson: 1836–1917* (Faber

and Faber, London, 1939); Ishbel Ross, *Child of Destiny: The Life Story of the First Woman Doctor* (Gollancz, London, 1950); E. Moberly Bell, *Storming the Citadel: The Rise of the Woman Doctor* (Constable, London, 1953); Jo Manton, *Elizabeth Garrett Anderson* (Methuen, London, 1965). The classic work on the American pioneers is Kate Campbell Hurd-Mead, *Medical Women of America: A Short History of the Pioneer Medical Women of America and of a few of their Colleagues in England* (Froben Press, New York, 1933).

5 The classic work here is Sheila Rowbottom, *Hidden from History: 300 Years of Women's Oppression and the Fight Against It* (Pluto Press, London, 1973).

6 See, for example, A. L. Wyman, 'The surgeoness: the female practitioner of surgery 1400–1800', *Medical History*, 28 (1984), pp. 22–41; Monica Green, 'Women's medical practice and health care in medieval Europe', *Signs*, 14 (1989), pp. 434–73. I say 'rediscovered', because women's medical practice in earlier centuries was well known to, for example, Sophia Jex-Blake; indeed, she used it as part of her argument for the admission of women to the medical profession. See her *Medical Women: A Thesis and a History*.

7 Jean Donnison, 'Medical women and lady midwives', *Women's Studies*, 3 (1976), pp. 229–50; *Midwives and Medical Men: A History of Inter-Professional Rivalry and Women's Rights* (Heinemann, London, 1977); Adrian Wilson, 'Participant or patient? Seventeenth century childbirth from the mother's point of view' in Roy Porter (ed.), *Patients and Practitioners: Lay Perceptions of Medicine in Pre-Industrial Society* (Cambridge University Press, Cambridge, 1985), pp. 129–44. On American midwives, see Jane B. Donegan, *Women and Men Midwives: Medicine, Morality and Misogyny in Early America* (Greenwood Press, Westport, Conn., 1978).

8 Barbara Ehrenreich and Dierdre English, *Witches, Midwives and Nurses: A History of Women Healers*, 2nd edn (Feminist Press, Old Westbury, NY, 1973); Mary Chamberlain, *Old Wives' Tales: Their History, Remedies and Spells* (Virago, London, 1981); Margaret Connor Versluysen, 'Old wives' tales? Women healers in English history' in Celia Davies (ed.), *Rewriting Nursing History* (Croom Helm, London, 1980); Green, 'Women's medical practice and health care'. See also Margaret Pelling and Charles Webster, 'Medical practitioners' in Charles Webster (ed.), *Health, Medicine and Mortality in the Sixteenth Century* (Cambridge University Press, Cambridge, 1979), pp. 165–235.

9 Rosemary White, *Social Change and the Development of the Nursing Profession: A Study of the Poor Law Nursing Service 1848–1948* (Henry Kimpton Publishers, London, 1978); Davies, *Re-writing Nursing History*; Christopher Maggs, *The Origins of General Nursing* (Croom Helm, London, 1983); Anne Summers, 'Pride and Prejudice: Ladies and Nurses in the Crimean

War', *History Workshop Journal*, 16 (1983), pp. 32–56; Monica E. Baly, *Florence Nightingale and the Nursing Legacy* (Croom Helm, London, 1986); Christopher Maggs (ed.), *Nursing History: The State of the Art* (Croom Helm, London, 1987); Anne Summers, *Angels and Citizens: British Women as Military Nurses 1854–1914* (Routledge and Kegan Paul, London, 1988); Robert Dingwall, Anne-Marie Rafferty and Charles Webster, *An Introduction to the Social History of Nursing* (Routledge and Kegan Paul, London, 1989).

10 The classic text of the Women's Health Movement is Boston Women's Health Collective, *Our Bodies Ourselves: A Health Book By and For Women* (Penguin, Harmondsworth, Middlesex, 1978; original USA edn, 1971). The classic text of 'old wives'' history is Ehrenreich and English, *Witches, Midwives and Nurses*. The other form of historical research inspired by the Women's Health movement has aimed at showing how medical categories and practices, especially in psychiatry and gynaecology, are aspects of men's oppression of women. See Ann Douglas Wood, ' "The Fashionable Diseases": Women's complaints and their treatment in nineteenth-century America' in Mary S. Hartman and Lois Banner (eds), *Clio's Consciousness Raised: New Perspectives on the History of Women* (Harper and Row, New York and London, 1974), pp. 1–22; J. L'Esperance, 'Doctors and women in nineteenth-century society: sexuality and role' in John Woodward and David Richards (eds), *Health Care and Popular Medicine in Nineteenth Century England: Essays in the Social History of Medicine* (Croom Helm, London, 1977); Barbara Ehrenreich and Deirdre English, *For Her Own Good: 150 Years of the Experts' Advice to Women* (Pluto Press, London, 1979).

11 Katherine Williams, 'From Sarah Gamp to Florence Nightingale: a critical study of hospital nursing systems from 1840 to 1897' in Davies (ed.), *Rewriting Nursing History*, pp. 41–75; Brian Abel-Smith, *A History of the Nursing Profession* (Heinemann, London, 1960), chs. 3–5.

12 E. Lutzker, 'Edith Pechey-Phipson M.D.: The Untold Story', *Medical History*, 11 (1967), pp. 41–5; Sandra L. Chaff, Ruth Haimbach, Carol Fenichel and Nina B. Woodside, *Women in Medicine: A Bibliography of the Literature on Women Physicians* (Scarecrow Press, New Jersey, 1977); Mary Ann Elston, 'Women's access to medical education in Great Britain, 1877– 1900: an overview', *Society for the Social History of Medicine Bulletin*, no. 41 (Dec. 1987), pp. 51–3. The new wave of research on American women doctors includes: Martha H. Verbrugge, 'Women in medicine in nineteenth century America', *Signs*, 1 (1976), pp. 957–72; Mary Roth Walsh, *Doctors Wanted, No Women Need Apply: Sexual Barriers in the Medical Profession, 1835–1975* (Yale University Press, New Haven, 1977); Dorothy Rosenthal Mandelbaum, 'Women in medicine', *Signs*, 4 (1978), pp. 136–45; Regina

Markell Morantz-Sanchez, *Sympathy and Science: Women Physicians in American Medicine* (Oxford University Press, New York, 1985).

13 The only study of the Ladies' Sanitary Association is W. C. Dowling, 'The Ladies' Sanitary Association and the Origin of the Health Visiting Service' (M.A. thesis, University of London, 1963). The corresponding health reform movement in the USA is described in Morantz-Sanchez, *Sympathy and Science*, ch. 2.

14 Baly, *Florence Nightingale and the Nursing Legacy*, esp. p. 22.

15 See, for example, her *Notes on Hospitals* (John W. Parker, London, 1859), being reprints of papers delivered at the NAPSS, evidence given to the Royal Commission on the State of the Army, and articles in *The Builder*.

16 Edward T. Cook, *The Life of Florence Nightingale* (2 vols, Macmillan, London, 1913); Cecil Woodham-Smith, *Florence Nightingale 1820–1910* (Constable, London, 1950); F. B. Smith, *Florence Nightingale: Reputation and Power* (Croom Helm, London, 1982); Charles E. Rosenberg, 'Florence Nightingale on contagion: the hospital as moral universe' in Charles E. Rosenberg (ed.), *Healing and History: Essays for George Rosen* (Dawson, Folkestone, 1979).

17 There is as yet no truly satisfactory survey of the sanitary reform movement. Recent works which deal with various aspects of it include Margaret Pelling, *Cholera, Fever and English Medicine 1825–1865* (Oxford University Press, Oxford, 1978); Rosenberg, 'Florence Nightingale on contagion'; F. B. Smith, *The People's Health 1830–1910* (Croom Helm, London, 1979); Anthony S. Wohl, *Endangered Lives: Public Health in Victorian Britain* (J. M. Dent and Sons, London, 1983); Christopher Hamlin, 'Providence and putrefaction: Victorian sanitarians and the natural theology of health and disease', *Victorian Studies*, 28 (1985), pp. 381–411; Virginia Smith, 'Physical Puritanism and sanitary science: material and immaterial beliefs in popular physiology, 1650–1840' in W. F. Bynum and Roy Porter (eds), *Medical Fringe and Medical Orthodoxy 1750–1850* (Croom Helm, London, 1987), pp. 174–97; Christopher Hamlin, 'Muddling in Bumbledom: on the enormity of large sanitary improvements in four British towns, 1855–1885', *Victorian Studies*, 31 (1988), pp. 55–83; Deirdre MacBean, 'The reception of germ theory in Britain' (M. Phil. dissertation, University of Cambridge, 1988).

18 Edwin Chadwick, *Report on the Sanitary Condition of the Labouring Population of Great Britain*, ed. M. W. Flinn (Edinburgh University Press, Edinburgh, 1965; original edn, 1842); Karl H. Metz, 'Social thought and social statistics in the early nineteenth century: the case of sanitary statistics in England', *International Review of Social History*, 29 (1984), pp. 254–73. On Chadwick's aims, see Julian Martin, 'Edwin Chadwick: social and medical reformer?' (M. Phil. essay, University of Cambridge, Wellcome

Unit for the History of Medicine, 1984). On the changed meaning of the word 'sanitary', see Pelling, *Cholera, Fever and English Medicine*, pp. 30–1.

19 *Transactions of the National Association for the Promotion of Social Science*, 1 (1857), p. xxvi (quoting from the minutes of the foundation meeting); Lawrence Goldman, 'The Social Science Association, 1857–1886: a context for mid-Victorian Liberalism', *English Historical Review*, 101 (1986), pp. 95–134; Ronald K. Huch, 'The National Association for the Promotion of Social Science: its contribution to Victorian health reform, 1857–1886', *Albion*, 17 (1985), pp. 279–300.

20 Mary Anne Baines, 'The "Ladies National Association for the Diffusion of Sanitary Knowledge"', *Transactions of the National Association for the Promotion of Social Science*, 2 (1858), p. 531. Bessie Rayner Parks, in 'The Ladies' Sanitary Association', *English Woman's Journal*, 3 (1859), pp. 82–3, used a very similar form of words, so they were probably both based on the Association's foundational documents.

21 Baines, ibid.

22 Ian Bradley, *The Call to Seriousness: The Evangelical Impact on the Victorians* (Jonathan Cape, London, 1976); Ford K. Brown, *Fathers of the Victorians: the Age of Wilberforce* (Cambridge University Press, Cambridge, 1961); Kathleen Heasman, *Evangelicals in Action: An Appraisal of Their Social Work in the Victorian Era* (Geoffrey Bles, London, 1962); Brian Harrison, 'Philanthropy and the Victorians', *Victorian Studies*, 9 (1965–6), pp. 353–74. Studies of evangelical or dissenting philanthropy with a specifically medical focus include Anne Digby, *Madness, Morality and Medicine: A Study of the York Retreat, 1796–1914* (Cambridge University Press, Cambridge, 1985); Robert Kilpatrick, ' "Living in the Light": dispensaries, philanthropy and fevers in late eighteenth century London' in Andrew Cunningham and Roger French (eds), *The Medical Enlightenment of the Eighteenth Century* (Cambridge University Press, Cambridge, 1990); Francis M. Lobo, 'John Haygarth, smallpox, and religious dissent in eighteenth century England' in Cunningham and French, ibid.; Amalie M. Kass and Edward H. Kass, *Perfecting the World: The Life and Times of Dr Thomas Hodgkin 1798–1866* (Harcourt Brace Jovanovich, Boston, San Diego, New York, 1988).

23 Susan R. Powers, 'The diffusion of sanitary knowledge', *Transactions of the National Association for the Promotion of Social Science*, 4 (1860), pp. 713–16; Dowling, 'The Ladies Sanitary Association', ch. 6; F. K. Prochaska, 'Body and soul: Bible nurses and the poor in Victorian London', *Historical Research*, 60 (1987), pp. 336–48.

24 Catherine Hall, 'The early formation of Victorian domestic ideology' in Sandra Burman (ed.), *Fit Work for Women* (Croom Helm, London, 1979), pp. 15–32; Leonore Davidoff and Catherine Hall, *Family Fortunes: Men*

and Women of the English Middle Class 1780–1850 (Hutchinson, London, 1987).

25 F. K. Prochaska, *Women and Philanthropy in Nineteenth Century England* (Clarendon Press, Oxford, 1980); Anne Summers, 'A Home from Home – women's philanthropic work in the nineteenth century' in Burman (ed.), *Fit Work for Women*, pp. 33–63; Jessica Gerard, 'Lady Bountiful: women of the landed classes and rural philanthropy', *Victorian Studies*, 30 (1987), pp. 184–209; Louis Billington and Rosamund Billington, ' "A Burning Zeal for Righteousness": women in the British anti-slavery movement, 1820–1860' in Jane Rendall (ed.), *Equal or Different: Women's Politics 1800–1914* (Basil Blackwell, Oxford, 1987), pp. 82–111; June Rose, *Elizabeth Fry* (Macmillan, London, 1980); Harriet Warm Schupf, 'Single women and social reform in mid-nineteenth century England: the case of Mary Carpenter', *Victorian Studies*, 17 (1973–4), pp. 301–17.

26 Barbara Leigh Smith, *Women and Work* (Bosworth and Harrison, London, 1857), p. 18.

27 Ibid., pp. 6–7.

28 Jane Rendall, ' "A Moral Engine"? Feminism, Liberalism and the *English Woman's Journal*' in Rendall, *Equal or Different*, p. 115.

29 Jane Rendall, 'Friendship and politics: Barbara Leigh Smith Bodichon (1827–91) and Bessie Rayner Parkes (1829–1925)' in Susan Mendus and Jane Rendall (eds), *Sexuality and Subordination: Interdisciplinary Studies of Gender in the Nineteenth Century* (Routledge, London, 1989), pp. 136–70; Rendall, ' "A Moral Engine"?', pp. 112–38. On the British women's movement and the *English Woman's Journal's* place in it, see Strachey, *The Cause*; Lee Holcombe, *Victorian Ladies at Work: Middle-Class Working Women in England and Wales 1850–1914* (David and Charles, Newton Abbot, 1973); Olive Banks, *Faces of Feminism: A Study of Feminism as a Social Movement* (Martin Robertson, Oxford, 1981); Sally Alexander (ed.), *Studies in the History of Feminism* (University of London, London, 1984).

30 Rendall, ' "A Moral Engine"?', pp. 114, 122, 126–8, etc.

31 Mrs Jameson, *The Communion of Labour: A Second Lecture on the Social Employments of Women* (Longman, Brown, Green, Longmans and Richards, London, 1856), pp. 22–4.

32 See, for example: Glen Petrie, *A Singular Iniquity: The Campaigns of Josephine Butler* (Macmillan, London, 1971); Judith R. Walkowitz, *Prostitution and Victorian Society: Women, Class and the State* (Cambridge University Press, Cambridge, 1980); Nancy Boyd, *Josephine Butler, Octavia Hill, Florence Nightingale: Three Victorian Women who Changed their World* (Macmillan, London, 1982); Coral Lansbury, *The Old Brown Dog: Women, Workers, and Vivisection in Edwardian England* (University of Wisconsin Press, Madison,

1985); Sheila Jeffreys, *The Spinster and her Enemies: Feminism and Sexuality 1880–1930* (Pandora Press, London, 1985); Mary Ann Elston, 'Women and Anti-Vivisection in Victorian England, 1870–1900' in Nicolas Rupke (ed.), *Vivisection in Historical Perspective* (Croom Helm, London, 1987), pp. 259–94.

33 *English Woman's Journal*, 3 (1859), p. 381. The office shared was at 14a Princes Street, the original home of the *Journal*.

34 Bessie Rayner Parkes, 'The Ladies' Sanitary Association', *English Woman's Journal*, 3 (1859), p. 82.

35 For other examples of the use of this tradition, see Andrew Cunningham, 'Thomas Sydenham: epidemics, experiment and the "Good Old Cause" ' in Roger French and Andrew Wear (eds), *The Medical Revolution of the Seventeenth Century*, pp. 164–90; Lobo, 'Haygarth, smallpox, and religious dissent'.

36 Pelling, *Cholera, Fever and English Medicine*, pp. 76–8, 105–6, etc.

37 See, for example, Chadwick, *Report on the Sanitary Conditions*, ed. Flinn, p. 214.

38 All the tract titles mentioned here are taken from the list in the Ladies' Sanitary Association Annual Reports for 1881 and 1882 (British Library shelfmarks CT 334 (9) and CT 291 (5) respectively).

39 Ladies' Sanitary Association Annual Report, 1882, p. 20.

40 See, for example, Ladies' Sanitary Association Annual Report, 1881, pp. 25–30.

41 Ladies' Sanitary Association Annual Report, 1882, p. 12, p. 11.

42 [Anon.] *The Bride's New Home* (Jarrold and Sons, Ladies' Sanitary Association, London, n.d.), p. 7. This and the other tracts quoted are in the Bessie Rayner Parkes collection at Girton College Library, Cambridge.

43 Ibid., pp. 12–13.

44 See, for example, S[usan] R[ugeley] P[owers], *How to Manage a Baby* (Jarrold and Sons, Ladies' Sanitary Association, London, n.d.).

45 *The Bride's New Home*, pp. 13–15.

46 Dowling, 'The Ladies' Sanitary Association', p. 131; see also Prochaska, 'Body and soul'; Summers, 'A Home from Home'.

47 Ladies' Sanitary Association Annual Report, 1881, pp. 31–2. Many less experienced members gave lectures based on Catherine M. Buckton's *Health in the House: Twenty-five Lectures on Elementary Physiology in its Application to the Daily Wants of Man and Animals* (Longmans, London, 1875).

48 *The Bride's New Home*, p. 7.

49 [Anon.], *Children Going to School* (Jarrold and Sons, Ladies' Sanitary Association, London, n.d.), p. 58. The other tracts were: *The Mother: Her Duty First Towards Her Unborn and Then to Her Newly-Born Infant*; *The*

Inspector: How to Get Rid of Bad Smells Without, and Bad Temper Within; *A Day in the Country*; and *Lost and Found*.

50 For an example of the latter, see Ladies' Sanitary Association Annual Report, 1881, p. 30.

51 Ladies' Sanitary Association Annual Report, 1882, p. 21.

52 [Anon.], *Village Work* (S. W. Partridge, Ladies' Sanitary Association, London, n.d.), pp. 13, 15, 16, 17, 18.

53 Celia Davies, 'The health visitor as mother's friend: a woman's place in public health, 1900–14', *Social History of Medicine*, 1 (1988), pp. 39–59, quotations at pp. 50–1. See also Dowling, 'The Ladies' Sanitary Association' ch. 8; Helen Jones, 'Women health workers: the case of the first women factory inspectors in Britain', *Social History of Medicine*, 1 (1988), pp. 165–81.

54 For example: 'It is not difficult to know / What works us hurt and harm; / God warns us by the sense of smell / When we should take alarm.' *Village Work*, p. 18.

55 Rosenberg, 'Florence Nightingale on contagion'; MacBean, 'The reception of germ theory in Britain'. Studies of the rise of germ theory in France include Bruno Latour, *The Pasteurization of France*, tr. Alan Sheridan and John Law (Harvard University Press, Cambridge, Mass., 1988; original edn, Metailie, Paris, 1984); Michael A. Osborne, 'French military epidemiology and the limits of the laboratory: the case of Louis-Felix-Achille Kelsch' in Andrew Cunningham and Perry Williams (eds.), *Medicine and the Laboratory* (Cambridge University Press, forthcoming).

56 See, for example, Hilary Rose, 'Hand, brain, and heart: a feminist epistemology for the natural sciences', *Signs*, 9 (1983), pp. 73–80; and 'Gendered reflections on the laboratory in medicine' in Cunningham and Williams (eds.), *Medicine and the Laboratory*.

3

Women in the Nineteenth-Century Scientific Instrument Trade

A. D. Morrison-Low

Skilled trade in the nineteenth century presents special problems for the historian. In the particular case of the scientific instrument trade, it is not just that women are 'hidden from history', but that not enough is known about this narrow economic area for generalizations to be made about the people, the conditions, outwork, pay, work practice, apprenticeship and so on. Therefore one is restricted to discussing particular cases. However, in itself this is a useful exercise, as it gathers together scattered details to present a partial picture, to be used with caution. This picture may be the only glimpse we can get into the lives of men and women whose work produced diverse instruments and apparatus, which at the least provided contemporary scientific entertainment and learning and at best furthered the frontiers of knowledge.

Unlike many nineteenth-century skilled trades which have been charted with greater or lesser success, the scientific instrument trade presents particular problems, which inhibit broad generalizations about its structure, its chronological development, and certainly about its day-to-day workings. Its work-force was never numerically large, nor, indeed, widespread; in fact, it was geographically limited to a few specific centres of population, until changing communications towards the end of the nineteenth century made this restriction unnecessary. It has left little documentary evidence concerning its practitioners, which means that we know almost nothing about the nitty-gritty of the running of the business, except in a few isolated cases, which are often unusual, and therefore cannot be treated as general examples. We know little of what 'instrument making' meant in terms of piece-work, outwork, homework, wholesaling, retailing, the buying in of parts, repairing, assembling or making from scratch. Nor do we know what were the implications of changes in technology, nor the introduction of the factory system. In fact, our understanding of the structure of the trade is severely limited, and has not been helped by an apparent inhibition on the part of researchers to proceed beyond the end of the previous (and possibly less complicated) century, into the largely uncharted regions of the nineteenth century. And if we know little of the men involved in the nineteenth-century instrument trade, we know still less about the women. It is essential, therefore, to refrain from all but the most tentative of generalizations, and after addressing the basic problems of what the instrument trade was, how it changed during the nineteenth century, and who its practitioners were, to devote most of this chapter to individuals, without claiming them as necessarily representative of their profession.

WHAT WAS THE SCIENTIFIC INSTRUMENT TRADE IN THE NINETEENTH CENTURY?

The origins of scientific instrument making are lost in antiquity;[1] and its development over the centuries has reflected humanity's interest in its surroundings and a desire to extend the human senses.[2] However, on a practical level, the instrument trade, in common with trade in other commodities, has had much to do with business acumen and the desire to make money. In brief, the modern beginnings of the trade are seen as a Renaissance activity, spreading gradually across Europe from Italy through the centre of the Continent. It eventually reached London, where it flourished during the eighteenth century, and became pre-eminent there in the Western hemisphere.[3] In the early nineteenth century, London was still the world centre of the trade,[4] but by mid-century this position had been lost. In part, this was because newer technology appeared later on the Continent, allowing French and German firms, with their more advanced plant, to produce items more cheaply and undercut the London craft-based industry, which clung to its time-honoured customs of production in small shops. But the near-demise of centralized London instrument trading was also a consequence of the rise of large wholesaling manufacturers, usually based in the English Midlands, whose access to railways allowed swift movement to domestic, as well as foreign, markets. Colonial markets, which for Britain expanded greatly during this period, were also important for their demand for surveying, mining and navigation instruments, but were not supplied exclusively by the mother country.

Within these markets, three broad groups of customers may be identified: the 'scientist', the 'professional' and the 'dilettante'.[5] What have come to be called 'scientific instruments' covered a wide variety of devices. These ranged from the new piece of observational or experimental apparatus, built to extend the frontiers of knowledge, often created by the instrument-maker in conjunction with the commissioning specialist (perhaps a surveyor, or an astronomer, or someone who would qualify for the late nineteenth-century description of 'scientist'). The term also included the working tools of various professions (the protractors and slide-rules of architects, the theodolites and levels of surveyors, the sextants and chronometers of navigators, the demonstration apparatus of teachers), to drawing-room toys, such as the kaleidoscope, and many other optical instruments, which merely amused the leisured upper classes.

Thus the trade catered for people engaged in a variety of activities, from the everyday to the occasional, from the extraordinary to the mundane. However, the trade itself was as diverse as its markets, and should not be identified as a single craft, but rather as a grouping of skills which reflected the range of required practical knowledge and abilities: from metal engravers to optical workers, from cabinet-makers to brass-tube drawers, from glass-blowers to ivory-turners. As a trade which had had its roots in craft organization, it seems to have been run along similar lines to other small-scale artisan producers: for instance, some of the metal trades of Birmingham,[6] and, in particular, that of gun manufacture.[7] Indeed, by mid-century, instrument manufacture was one of the industries which had begun to move out of London and into the hands of businessmen and entrepreneurs, such as the wholesaler James Parkes & Son of Birmingham, or J. P. Cutts and Chadburn Brothers, both of Sheffield: this is demonstrated by the volume and scope of each firm's trade literature. R. H. Nuttall explains this removal in terms of the demand for skilled mechanics needed to construct and install machinery in the rapidly-expanding cotton industry; and goes on to comment that in London:

many of the makers to the trade [that is, the subcontractors] must have led a hand-to-mouth existence of total dependence upon the retailer; for they were tied to him for orders, for the components of the instruments they

made, and for payment. Such a man was a near-captive; to become a retailer
he required finance few could aspire to; though for those who were
successful in acquiring retail premises there accrued increased opportunities
for earnings, and perhaps improved status. On the other hand the short
working lives of many of those opticians and instrument makers who appear
in the trade directories, is indicative of a high likelihood of failure. To such
men the prospect of alternative, better paid employment in the new
industrial areas must have been welcome.[8]

This wholesaling trade was not exclusive to the Midlands: J. J. Hicks of
London was perhaps the most prominent of the London retailers, judging
by the material available in his catalogues between 1875 and 1913.[9]
However, of the organization of such businesses, in which many obscure
makers to the trade may have been working in their own or wholesalers'
premises, working on parts or finishing pieces, and using their own or
their masters' tools, virtually nothing is known. Nor are the nature or rate
of change brought about by industrialization understood.[10]

Instruments which have an engraved 'maker's' name were often the end
product of the labour of a number of people, possibly not all working
under one roof; and the signature of the 'maker' may well have been the
retailer or wholesaler rather than the manufacturer. Other businesses
bought in complete instruments, and engraved their own names on them;
for instance, T. N. Clarke and others have demonstrated that a substantial
number of Scottish firms were buying in items from English wholesalers,
especially the more cheaply manufactured items made for 'dilettante'
consumers, such as microscopes and telescopes, which could be produced
in large batches.[11] Later in the century, when French and German
microscopes were produced more cheaply than in Britain, some British
opticians sold continental pieces, often with both the maker's and the
retailer's names.[12]

A recent study by Lucio Sponza looks at another group within the
British instrument-making trade: the Italians. There was a particularly
large Italian immigrant community of artisans in the Holborn district of
London, many of whom made a living as barometer- and thermometer-
makers. Often they married English wives, and these were to be found
occasionally helping out in the business as 'French polishers'. But, as
Sponza comments: 'The concept of "working women" was (as it generally
still is, a century later) only associated with activities outside the domestic
sphere; any partnership between husband and wife in running a small

shop was regarded as a real job for the men, and a mere secondary and supportive role for the women, whatever their commitment.'[13]

WHO WERE THE MAKERS?

The involvement of women in the manufacturing trades has a history which predates the nineteenth century, and their role has been examined to some extent by various authors. It would appear that even in the seventeenth century women were not excluded completely from crafts or trades, and that while unmarried women had little or no economic status, the position of married women was that they assisted their husbands, and could assume their positions in business upon widowhood, subject to certain restrictions.[14] Ivy Pinchbeck claims that by the late eighteenth century 'the craftsman's wife was usually so well acquainted with her husband's business as to be "mistress of the managing part of it," and she could therefore carry on in his absence or after his death, although she herself might lack technical skill.'[15] Indeed, in technical trades, 'a widow usually engaged able workmen to assist her, whilst retaining the management in her own hands.'[16] For nineteenth-century London, Sally Alexander produces a model of the 'sexual division of labour . . . sustained by ideology not biology, an ideology whose material manifestation is embodied and reproduced within the family and then transferred from the family into social production.' However, she does not discuss the artisan classes, possibly because skilled labour does not fit her model. Yet her discussion of unskilled women's work remains a valid portrait of those uncounted women who must have spent hours in the home doing piece-work.[17] Geoffrey Crossick provides a picture in which 'industrialization in Britain produced no withering away of small-scale production', providing possibilities for small producers, particularly in the consumer trades, of which the instrument trade was merely one.[18] Against this complex and busy background, the instrument-making trade changed over the nineteenth century from largely London-based crafts into a number of diverse businesses which had the potential to find international markets from such apparently unlikely places as Dublin and Cambridge.[19]

Numerically, the size of the trade was not large; and certainly compared with other skilled areas it was very small indeed. Willem Hackmann has used figures taken from the census returns between 1841 and 1881, as analysed for occupation by Charles Booth, which were checked and extended by W. A. Armstrong to include 1891, to enable him to assess the probable size of the work-force. He has, however, pointed out that the figures do not include retailers, nor those who may also have worked in other categories, for example, as cabinet-makers.[20] So small is the enumerated work-force that it can be measured in thousands at most, and the female component of this in hundreds (see appendix). Although it is often illuminating to look at census returns to see what the individual enumerator recorded, details vary with each census, and much depended on the diligence and inclination of the writers.[21]

Of course, much also depended on what the individual head of household was prepared to reveal to the enumerator, and, as is the case today with small family businesses, often the wife's role is hidden until she is forced through her husband's death or bankruptcy to assume a more visible one. Perhaps an important part of the work of an instrument-maker's wife was to 'keep the books'; actually, to run or maintain the finances of the business for which her husband made or assembled the stock. Trading could continue with the widow employing journeymen after her husband's death, and certainly in an earlier period, when the guild structure in London was more effective, a widow was held responsible for the training of existing and new apprentices. M. A. Crawforth has commented that there

> are several implications consequent to the succession of a widow. It suggests that her husband had been running a business in which she had been involved to such an extent that she could continue the business. The training of apprentices infers the employment of at least one journeyman with formal training in the industry, and the approval of the guild officials in binding an apprentice to a widow indicates their confidence that she could reasonably be expected to continue in business for at least seven years.[22]

Widowhood could bring with it privileges not otherwise accessible to single or married women: property rights, the trusteeship of children and apprentices, and responsibilities which would otherwise have been exclusively male.

Perhaps the most readily-available source for revealing the identity of the women involved in the scientific instrument trade is the street directories, which covered the major centres of population, and began appearing annually towards the end of the eighteenth century.[23] Outside

London, smaller centres of the trade had been in existence within Britain and Ireland for some time: for instance, in Dublin and in Cork since the late seventeenth century,[24] in the Scottish towns of Edinburgh, Glasgow, Aberdeen, Dundee and Greenock from a similar date,[25] and also in the English provincial centres of Manchester,[26] Liverpool, and, later, Birmingham and Sheffield.[27] However, at the beginning of the nineteenth century, the greatest concentration of the trade was in the small workshops in Clerkenwell, London, fringed by those in the City.[28]

The London street directories reveal over the years the small numbers of women who ran firms which sold scientific instruments: for instance, Mrs Eliza Finch, drawing instrument maker, 30 Ranwell Street, Bow, between 1886 and 1895;[29] or Miss Louise S. Grosvenor, in business as a mathematical instrument maker for 1855 and 1856.[30] Penelope Steel, widow of David Steel, ran a chart-selling and navigation warehouse at Little Tower Hill from 1803 to 1805.[31] Mrs Elizabeth Irvine, widow of John Irvine, was the head of an optician's shop in Hatton Garden from 1838 to 1846.[32] Mary Wellington, widow of Alexander Wellington, was in business as a mathematical instrument maker with her son between 1813 and 1815, and as an optician in her own name, alone, at Crown Court, Soho, from 1816 to 1827.[33] Charlotte Rubidge, widow of James Rubidge, continued his scale-making business from 1820 to 1827 in her own name, at 18 Great Eastcheap.[34] Mrs William Holyman, nautical- and mathematical instrument manufacturer, continued her late husband's business off Commercial Road from 1860 to 1874.[35] Widow Stretton, mathematical instrument maker, Rotherhithe, London, was recorded between 1818 and 1824.[36] Ann Bradberry, the widow of Robert Bradberry, carried on his business as a patent spectacle-maker at 27 Hollis Street, Cavendish Square, from 1820 to 1827, when the firm became Bradberry & Co.[37]

A few women appeared in business for only very short periods, possibly in order to wind down a firm, or because they got into difficulties and had to sell up: for instance, Mrs Ellen Lowe, widow of Joshua Lowe, appeared for a single year as an optician in Borough for 1849;[38] likewise, Elizabeth Taylor, widow of John Taylor, was recorded as an optician only in 1830 and 1831;[39] and in her survey of instrument makers, E. G. R. Taylor recorded three women, Hannah Doublott, Clara Davis and Mary Hubee, each described as optician, mathematical and philosophical instrument maker, for the year 1834 only.[40] Mrs Mary Bracher, the widow of George Bracher, working optician and brass founder, ran her late husband's business in Commercial Road East for two years before it was sold to Thomas Reed, optician, in 1845;[41] similarly, the instrument making business of Henry Nelson's widow Louisa, at 2 Gloucester Street, Queen's Square, lasted only from 1847 until 1850, when Thomas Mason, drawing instrument maker, was recorded at this address.[42] Mrs Louisa Thompson, widow of Joseph Barry Thompson, a mathematical instrument maker recorded between 1827 and 1848, only appears in 1849 and 1850, before apparently selling up to James Croger, optician.[43] Mrs Sarah Hudson, widow of John Hudson, optician, lasted from 1838 to 1842 at 17 Ryder's Court, Leicester Square, before selling the business to Sargeant Allen, optician.[44] Mrs Mary Phelps, widow of Thomas Phelps, managed to stay in business as an optician between 1841 and 1845, when the name changed to James Phelps, presumably a son.[45] The nomination of a wife as executrix both confirmed her ability, and made her responsible for winding up or continuing business, dealing with stock, orders, work in hand, and assets, such as patents.

However, there were also long-lasting female business successes, such as Mary Jones, widow of Owen Jones, who continued her late husband's business at 241 Oxford Street as optician, philosophical and mathematical instrument maker for thirteen years from 1828 to 1841, before selling up to William Dixey;[46] Mary Ann Holmes, widow of John Holmes, was described as a spectacle-maker and optician between 1831 and 1838;[47] Miss Neriah Snart, who appears to be, unusually, the daughter rather than the widow of John Snart, was listed as an optician, and mathematical and philosophical instrument maker for nine years between 1838 and 1847. She was followed by Miss Martha Snart, who ran the business between 1848 and 1854.[48] Even the four-year business of Mrs Sarah Hillum, widow of Richard Hillum, optician and mathematical instrument maker between 1849 and 1853, before the business became Hillum & Co., compares well with the success of other similar businesses at the time, whether run by men or women, if time in business alone is considered an index of success.[49] In 1851 Mrs Harriet Mills took over her late husband's specialist telescope making business, first recorded in 1824, and ran it until 1869, when it was bought up by another telescope maker, Alexander Clarkson.[50] Clarke and others have demonstrated that upon closer inspection the instrument trade was subject to great insecurity, possibly because of the amount of capital which was tied up in stock on the premises or in special commissions, and that even famous and apparently economically successful firms went bankrupt. For instance, the Glasgow business run by James White, which later went on to become the sole supplier of instrumentation devised by Lord Kelvin, and included lucrative Admiralty contracts for navigation compasses, as well as electrical instruments and apparatus for submarine telegraphy, was sequestrated in 1861.[51]

It is seldom known why individual women went into business, but it would appear from most of the examples found in the street directories that widows would carry on their late husband's firm, probably because they were already involved and could offer some form of continuity, sometimes to allow a young son time to grow up and enter the trade. One such example is that of Elizabeth Rabone, née Smith, described as a rule-maker, of Snowhill, Birmingham, widow of Michael Rabone, who died about 1803; their son John took over the business in 1817 at the age of twenty-two. During his mother's trusteeship, the firm was called Elizabeth Rabone until 1808, and Elizabeth Rabone & Son from then until 1817. Even so, Mrs Rabone continued to be listed as a rule-maker until 1835, so she must have retained an interest in the firm's affairs after her son assumed his father's position as head of the business.[52] Similarly, Mary De Grave, a London scale-maker, widow of Charles, continued his business in her own name after his death from 1800 to 1816, and between 1817 and 1844 it was described as Mary De Grave & Son, before becoming De Grave, Short & Fanner.[53] Elizabeth Marriott Bardin, listed as a globe-maker from 1821 to 1859, must have had family connections, as yet uncovered, with the earlier London globe-makers W. & T. M. Bardin, and Thomas Bardin, who were at the same address.[54]

There are further similar examples to be found in London: Mrs Agnes Syeds, nautical instrument and compass maker in her own name between 1834 and 1853 at 379 Rotherhithe; she was the widow of John Syeds, active from 1790 to 1810, followed by the partnership of Syeds & Davis until 1833. From 1853 until 1864 the business was run by John Ramsey Syeds, presumably a relative.[55] Mrs A. Southouse, described as a nautical instrument maker at 25 Goodman's Yard, Minories, succeeded J. Southouse in 1867, and was succeeded by Robert Southouse in 1870.[56] Mrs Emily Tree, rule-maker, replaced James Tree at 22 Charlotte Street, Blackfriars Road, in 1843, and was herself replaced in 1852 by James Tree & Co.[57]

The family was the key to these women's involvement in business. As Davidoff and Hall have convincingly demonstrated, middle-class and lower middle-class women were involved in economic enterprise with their husbands during this period to an extent that can only be uncovered by looking at many particular examples in some detail.[58] Elsewhere, the class has been defined as the 'petite bourgeoisie' by Crossick and Haupt, who stress that

> the commonplace necessities of family ties (not just marriage, but the occupations of brothers and sisters, of sons and daughters), personal mobility, and the flow of individuals between wage employment and independent enterprise, meant that the relationship between working class and petite bourgeoisie was a real and lived aspect of both daily and lifetime experience for much of the urban population of nineteenth-century Europe.[59]

Mary and Ann Dicas were both daughters of John Dicas, who was a Liverpool liquor merchant. His hydrometer, a device for measuring the specific gravity of liquids, and therefore of interest to the Excise, was patented in England in 1780,[60] and adopted that year by the United States government for estimating the strength of imported liquors. In London, a Board of Inquiry of the Royal Society was set up in 1802 to investigate the rival merits of a number of different hydrometers then in use throughout the British Isles, with a view to choosing the most effective for revenue use by the Excise. Because her father was deceased by this time, Mary Dicas submitted his instrument, and travelled to London to explain its principles to the board.[61] She must, therefore, have had some degree of technical understanding. She appears in the street directories as a hydrometer maker from 1797 to 1806, and the following year with George Arstall, a scale beam maker, whom she married.[62] Ann Dicas, her sister, appears as a patent hydrometer maker between 1818 and 1821, and after her marriage to Benjamin Gammage in May 1821, he became the 'only proprietor of the patent'.[63]

The work of this same Board of Inquiry illustrates another way in which family connections, through women, exerted influence, and ultimately won lucrative contracts. Another hydrometer which was presented to the board for inspection in the hope of being chosen, was that devised by Bartholomew Sikes, who worked in the Excise Department for almost fifty years. He also produced a set of specialized excise tables, which greatly impressed the committee. However, he did not live to reap his reward for this work, as he died, aged seventy-three, in October 1803. His widow petitioned for the approval of her late husband's hydrometer in May 1805, and in December 1806 suggested £3,000 for the rights to the instrument. The Revenue thought this over-valued, and in January 1807, Mary Sikes (now remarried) reduced the sum to £2,000, but proposed that Robert Brettell Bate, her nephew and son-in-law, should be granted the sole right to manufacture the instruments.[64] Although the Act did not come into force until 1818, R. B. Bate nevertheless won the contract, and subsequently managed to win other government contracts (such as being appointed the sole manufacturer of the imperial weights and measures introduced by the Act of 1824).[65] After Bate's death, in late 1847, his widow petitioned the Excise to continue 'to supply your Honorable Board with Gauging Instruments and the Hydrometers and Saccharometers invented by her deceased Father and Husband' (Bate had invented a new form of saccharometer, sanctioned for exclusive use by the Excise after 1829);[66] and the request was granted; but after a couple of years she decided to retire from business, and none of her children was prepared to continue the firm. A petition from another firm of specialist hydrometer-makers, Dring & Fage, applying for the position left vacant, records unequivocally that

the circumstance of an improvement having been introduced in the Hydrometer by a Mr Sikes, . . . was Ordered by Parliament to be the instrument used in taking the Revenue. For his invention Mr Sikes was Compensated by a Government Grant of a considerable sum of money – Subsequently to which time the Family obtained as a further remuneration the exclusive manufacture of the instrument for the use of the Revenue, and Mr Bate being a relation by Marriage was selected to carry the business on for the benefit of the Family – That the Family have now enjoyed this privilege upwards of 30 years and the only remaining family representative is the Widow of Mr Bate, who determines to resign the Business, and retires on a comfortable competence and the original intention of the Government Grant has now been fully accomplished.[67]

It is perhaps worth noting that Dring & Fage, too, were a family concern; and that their business succession on at least three occasions went to wives or daughters, and through them to their husbands' families.[68]

Apprentices who married their masters' daughters could find themselves inheriting a profitable business; but while it is fair to say that marriage was considered more of an economic contract than it is today, it is not possible to say whether it was entirely loveless.[69] In the eighteenth century there is the example of Jesse Ramsden, who married Peter Dollond's sister, Sarah, and received a share in the achromatic objective patent as part of her dowry,[70] and there must have been other, less famous examples. Hugh Powell (1799–1883), who became one of London's most renowned specialist microscope makers, initially worked with Peter Lealand senior, a skilled instrument maker; Powell married his daughter, and in 1841 took his brother-in-law, Peter Henry Lealand, into partnership. The business remained a small, highly-respected, family concern well into the twentieth century.[71] Another contemporary optical instrument maker, Andrew Ross, whose microscopes were also highly regarded, died in 1859, and the business was split between his son, Thomas, who inherited the microscope side, and his son-in-law, John Henry Dallmeyer, who succeeded to the camera business.[72] A Scottish example, from earlier in the century, is that of David Heron's daughter Jessie, who married James Whyte, an instrument maker who eventually rescued his master's, and her father's, ailing business after sequestration and turned it into a thriving concern.[73] Two immigrant Italian barometer-making families, that of Casartelli, based in Liverpool, and Ronchetti, based in Manchester, appear to have intermarried in at least two generations, with implications for the inheritance of their businesses.[74] Thus, through the act of marriage, it is possible that women played a central, but invisible, role in cementing and extenuating a close-knit scientific and business comunity, and ensured the cultural reproduction of trade knowledge.

These women, the wives and daughters of instrument makers and sellers, would not necessarily figure by occupation in any census return, as their role in the day-to-day running of the business is unclear. Other than keeping the books, it is quite possible that what they did was relatively menial and unskilled; Sponza suggests that women's role was generally supportive rather than definitive.

At an earlier period, a small number of girl apprentices were bound, through the London guilds, to instrument makers. M. A. Crawforth feels that it was unlikely that these examples represent women who were trained in the art of instrument making, particularly as the period of binding was very short: 'It is necessary to remember, however, that some workshop activities are equally well, or better, done by women than by men, such as fine filing, polishing, lacquering, and stringing of balances.'[75] Such a situation is illustrated in a late nineteenth-century account of James White's Glasgow workshops. In terms of the career prospects for a man:

having obtained such a situation, a boy soon finds he has his work cut out for him to become a fully-fledged journeyman optician, as instrument makers are pleased to style themselves . . . After some time at the vice, an apprentice will, perhaps, be sent to make terminals, or screws, or other small turned parts, in quantities, after which he will finish his seven years' apprenticeship working under various journeymen, making and finishing complete instruments . . . Altogether it is an extremely interesting occupation, and one which any thinking lad can never weary of.

The same account mentions the 'small shop full of girls, working chiefly in glass tube', and notes that 'a very delicate operation of the glass tube workers is drawing capillary tube for the syphon recorder', a sophisticated telegraphic instrument. Women were also employed in the Coil Winding Department, where 'an expert staff of girls daily convert miles of silk-covered, or otherwise insulated, wire into coils of every size and every shape.' The descriptions of these repetitive, yet dexterous, chores contrast strongly with the 'extremely interesting occupation' enjoyed by the 'thinking lad' of the same account.[76]

It is probably true to say that most of the women uncovered by the nineteenth-century censuses who worked in the instrument trade were the anonymous hundreds who drew capillary tube or wound wire coil; those who were not enumerated would be those who worked alongside their husbands in small workshops, possibly even at home, polishing, filing or finishing piece-work. As Duncan Bythell has succinctly put it:

> outwork did not die with industrialisation, it merely changed its forms: it disappeared in one area only to crop up in another. Indeed it is probably indestructible, for the circumstances which kept it going in the past show no signs of vanishing: so long as there are women who find it hard to make ends meet as they strive to run their homes, there exists a cheap and docile labour force for anyone who wishes to use it; the only requirement is that the work provided should be such as can conveniently be done in the home.[77]

The women who ran instrument businesses are therefore the exceptions, in that they are visible at all; and although it is possible to uncover in some detail the outlines of their working lives, there is much that can only be guessed. The perfunctory character of trade directories may even lead us to underestimate a woman's scientific acumen. For instance, according to the London trade directories, Mrs Janet Taylor (1804–70) ran a nautical academy and chart warehouse, and sold nautical instruments at 104

Minories between 1846 and 1874, in succession to her husband, George Taylor, who was active from 1837 to 1845, the firm becoming Mrs Janet Taylor & Co.[78] Yet further research has revealed that Mrs Taylor was the moving entrepreneurial spirit in this venture from the start, despite it being run in her husband's name; that she had been taught navigation by her father, a County Durham clergyman; that she was an authoress of a number of navigation texts; that she took out a patent in 1834 for her 'Mariner's Calculator'; and that she employed William Reynolds from about 1845 to make a range of nautical instruments.[79] Examples of these were exhibited in Mrs Taylor's name at both the Great Exhibition of 1851 and the International Exhibition of 1862.[80] Compasses, sextants and other nautical instruments bearing her name survive in public collections.[81] It is interesting to note, in passing, that there must have been similar practices in Continental instrument workshops, as widows displayed instruments at various international exhibitions.[82]

However, at least one Frenchwoman in particular seems to have had a long and successful career as an instrument-maker, and was credited in her husband's obituary in 1863 as responsible for the prolonged use of the dividing engine which he had constructed: 'C'est vers cette époque, qu'en dehors des travaux qui lui étaient commandés, il commença à construire la machine qui, pendant de longues années entre les mains de madame Brunner, devait diviser un si grand nombre de cercles sortis des ateliers de ses confrères.'[83] The pre-eminent Parisian firm of Brunner flourished between about 1835 and 1895; Johann Brunner was born in Switzerland in 1804, but the facts about his wife have still to be uncovered.

Two Scottish examples show that other women were more involved in the economic side than the creative part of the business. Mrs Margaret Gardner, widow of John Gardner junior, can be identified as the 'M. Gardner' of M. Gardner & Sons, which traded in Glasgow between 1823 and 1836. Bankruptcy in 1832 forced them to sell their Bell Street business, although they managed to acquire premises elsewhere and continue trading; the papers covering this distressing event show that although there was no formal deed of copartnery, Margaret was as much involved as her two sons, then aged twenty-three and twenty-seven.[84] Mrs Eliza Lennie ran her late husband's optical business in Princes Street, Edinburgh, from his death in 1854, and in 1872 married James Taylor, an ophthalmic optician from Haywood, England, who worked at the same premises under his own name between 1859 and 1887; it is unclear when Mrs Lennie retired from the business, allowing her first husband's optician children to take over. However, it appears that most of the instruments sold with a 'Lennie' signature were bought in from Continental wholesalers, and that the firm's practical expertise lay in ophthalmic work.[85]

At an earlier date, in 1805, the widow of Angelo Lovi, glass-blower, who continued her late husband's business in Edinburgh from 1805 to 1827, took out a patent with an advocate named J. R. Irving for 'Apparatus for determining the specific gravity of fluid bodies', involving developments to the hydrostatic bubbles invented by Alexander Wilson in the mid–1750s.[86] However, most women involved in the glass-blowing side of instrument-making – that is, making thermometers and barometers – appear to have been the widows of craftsmen, and businesswomen rather than entrepreneurs like Mrs Lovi.[87]

PLATE 3 Mrs Lovi's Patent Aerometrical Beads. A large set of the 'Patent Aerometrical Beads', with sliding rule and two editions of the instructions for their use. The apparatus was patented by Mrs Isabella Lovi and J. R. Irving in 1805.

(Reproduced by kind permission of the Trustees of the National Museums of Scotland)

SUMMARY

Women were evidently much involved in the nineteenth-century instrument trade. Their roles, however, varied immensely. The most readily identifiable women were those running a business, since they were listed in trade directories. But often they took over only when widowed, whilst a youthful son and heir completed his apprenticeship. The unseen role that they may have played before their husbands' deaths is difficult to assess, but if they had a long-term involvement in the financial side of the business this would certainly have enabled them to run the firm successfully when widowed. The succession of running a business, if there was no apparent male heir, might pass through a female heir to her husband's family. A number of instruments engraved with an apparent woman 'maker's' name survive, but whether the woman in question ever turned at the lathe, or fitted or adjusted the optical parts, is a matter for speculation. Other women, less readily identified by name, drew glass, worked in the cabinet shops, and generally undertook less skilled work. Only with further investigation into the nature of the trade and its changing characteristics will we discover more.

APPENDIX A

The following tables are taken from Charles Booth, 'Occupations of the people of the United Kingdom 1801–81', *Journal of the Statistical Society*, 49 (1886). The figures are for people in the stated groups, employed in the single classified entry for watches, instruments and toys. Numbers are given in units of 1,000, and to only one decimal place.

ENGLAND AND WALES

	1841	1851	1861	1871	1881
Males −15	{ 2.8	0.8	1.1	0.9	0.8
15−20	{	3.4	4.6	5.0	5.7
20−25	{	3.0	5.2	4.6	6.2
25−65	{16.8	15.2	20.3	23.4	27.0
65+	{	1.1	1.6	2.0	2.0
Total	19.6	23.5	32.8	35.9	41.7
Females					
−15	{ 0.8	0.1	0.1	0.1	0.1
15+	{	1.2	2.8	2.9	3.3
Total	0.8	1.3	2.9	3.0	3.4
Totals	20.4	24.8	35.7	38.9	45.1

Numbers [and percentages] of employed (in 1,000s):
20 [0.3], 25 [0.3], 36 [0.4], 39 [0.4], 45 [0.3]

SCOTLAND

	1841	1851	1861	1871	1881
Males −15	{ 0.2	0.1	0.1	0.1	0.1
15−20	{	0.4	0.4	0.6	0.8
20−25	{	0.2	0.4	0.4	0.7
25−65	{ 1.4	1.1	1.6	1.7	1.9
65+	{	0.1	0.1	0.1	0.1
Total	1.6	1.9	2.6	2.9	3.6
Females					
−15	−	−	−	−	−
15+	−	−	0.1	0.1	0.2
Total	−	−	0.1	0.1	0.2
Totals	1.6	1.9	2.7	3.0	3.8

Numbers [and percentages] of employed (in 1,000s):
2 [0.2], 2 [0.1], 3 [0.2], 3 [0.2], 4 [0.2]

IRELAND

	1841	1851	1861	1871	1881
Males −15	–	–	–	–	–
15−20	{		0.2	0.2	0.1
20−25	{ 1.1	1.1	0.1	0.1	0.2
25−65	{		0.9	1.0	1.0
65+	{		0.1	0.1	0.1
Total	1.1	1.1	1.4	1.5	1.4
Females					
−15	–	–	–	–	–
15+	–	–	0.1	0.1	–
Total	–	–	0.1	0.1	–
Totals:	1.1	1.1	1.4	1.5	1.4

Numbers [and percentages] of employed (in 1,000s):
1 [−], 1 [−], 2 [0.1], 1 [−]

APPENDIX B

The following tables are taken from W. A. Armstrong, 'The use of information about occupation: Part 2. An industrial classification 1841–1891' in E. A. Wrigley (ed.), *Nineteenth-Century Society, Essays in the Use of Quantitative Methods for the Study of Social Data* (Cambridge University Press, Cambridge, 1972), p. 268: figures are expressed in decimals of 1,000.

WATCH- AND CLOCK-MAKING

	1841	1851	1861	1871	1881	1891
Male	13.3	17.1	20.7	20.7	22.6	22.5
Female	0.2	–	0.6	0.6	0.8	1.4
Total	13.5	1.1	20.8	21.3	23.4	23.9

PHILOSOPHICAL AND SURGICAL INSTRUMENT AND ELECTRICAL APPARATUS MAKERS

	1841	1851	1861	1871	1881	1891
Male	2.1	2.9	3.9	4.5	7.1	18.1
Female	0.1	0.5	0.4	0.5	0.5	1.4
Total	2.2	3.4	4.3	5.0	7.6	19.5

ACKNOWLEDGEMENTS

I would like to thank D. J. Bryden, Dr Anita McConnell, Dr A. D. C. Simpson and Dr Mari Williams for their helpful comments on drafts of this paper, and Ken Smith for photography.

NOTES

1 The Antikythera Mechanism is usually regarded as the earliest extant scientific instrument. This Greek calendrical device was recovered from a wreck in 1900, and dated to the first century BC. See D. J. de S. Price, 'Gears from the Greeks', *Transactions of the American Philosophical Society*, new series, 64 (1974), pp. 1–70. A somewhat later piece of gearing, dating from between AD 480 and 560, surfaced in 1983, and is discussed by J. V. Field and M. T. Wright, 'Gears from the Byzantines: a portable sundial with calendrical gearing', *Annals of Science*, 42 (1985), pp. 87–138. Both these items, and how they relate to the development of Western timekeeping devices, are discussed in J. V. Field and M. T. Wright, *Early Gearing, Geared Mechanisms in the Ancient and Medieval World* (Science Museum, London, 1985).

2 A recent discussion of the literature of the history of scientific instrumentation is given by Willem D. Hackmann, 'Instrumentation in the theory and practice of science; scientific instruments as evidence and as aid to discovery', *Annali dell'Istituto e Museo di Storia della Scienza di Firenze*, 10 (1985), pp. 87–115.

3 For a recent discussion, see A. J. Turner, *Early Scientific Instruments 1400–1800* (Sotheby's Publications, London, 1985), pp. 171–230.

4 The reasons for this are explored by Willem Hackmann, 'The nineteenth-century trade in natural philosophy instruments in Britain' in P. R. de Clercq

(ed.), *Nineteenth-Century Scientific Instruments and Their Makers* (Rodopi, Leiden and Amsterdam, 1985), pp. 53–91.

5 D. J. Bryden, *Scottish Scientific Instrument-Makers 1600–1900* (Royal Scottish Museum, Edinburgh, 1972), pp. 7–23.

6 In general, although there were exceptions, instrument making was not included as a category amongst the Birmingham toy trades, but regarded as a more skilled trade. For a recent discussion on the metal and hardware trades, and the Birmingham toy trades, see Maxine Berg, *The Age of Manufactures 1700–1820* (Fontana, London, 1985), pp. 264–314.

7 See Clive Behagg, 'Masters and manufacturers: social values and the smaller unit of production in Birmingham, 1800–50' in Geoffrey Crossick and Heinz-Gerhard Haupt (eds), *Shopkeepers and Master Artisans in Nineteenth-Century Europe* (Methuen, London, 1986), p. 146: 'The major part of the work was subcontracted to small masters, and industrial expansion only served to extend the process of decentralized production. Gun barrels, forged and rolled by steam in large factories since the turn of the century increased the demand for the smaller units of production which generally carried out separately the nine processes involved in the completion of the gun. There was a strong craft basis here perhaps reflected best in the continuing insignificance of female and juvenile labour. In 1841 women and girls made up only 3.6 per cent of the total gun-making workforce. By 1851 this had only risen to 4 per cent. The figures for a more mechanized trade like button-making would have been closer to 40 per cent in 1841 and 57 per cent in 1851.'

8 R. H. Nuttall, *Microscopes from the Frank Collection 1800–1860* (A. Frank, Jersey, 1979), 'The Instrument Trade', pp. 8–13.

9 See, for example, J. J. Hicks's trade catalogues: *Illustrated and Descriptive Catalogue of Standard, Self-recording and other Meteorological Instruments, and all kinds of Chemical and Philosophical Apparatus Manufactured by James J. Hicks* (London, n.d.[c.1875]); *Illustrated Price List of Brewers Thermometers* (London, 1891); *Illustrated Price List of Aneroid Barometers, Microscopes, Sunshine Recorders, etc., Manufactured by James J. Hicks* (London, 1892); *Illustrated Price List of Aneroid Barometers, Standard Mercurial Barometers, Standard Thermometers, Sunshine Recorders, Anemometers, Rain Gauges and Meteorological Instruments of every description, Surveying and Other Instruments manufactured by James J. Hicks* (London, 1913). I am grateful to John Burnett for these references.

10 However, one 1850 account is as follows: 'I have known the camera obscura business for twenty-five years or so; but I can turn my hand to clock-making, or anything. My father was an optician, employing many men, and was burnt out; but the introduction of steam machinery has materially affected the optical glass grinder – which was my trade at first. In a steam-mill in

Sheffield, one man and two boys can now do the work that kept sixty men going': E. P. Thompson and Eileen Yeo (eds), *The Unknown Mayhew: Selections from the Morning Chronicle 1849–50* (Penguin, Harmondsworth, 1973), p. 355.

11 T. N. Clarke, A. D. Morrison-Low and A. D. C. Simpson, *Brass & Glass: Scientific Instrument Making Workshops in Scotland as Illustrated by Instruments from the Frank Collection at the Royal Museum of Scotland* (National Museums of Scotland, Edinburgh, 1989), pp. 73, 74, 83, 116, 163, 179, 187, 248, 253, 271 and 293 for instruments probably manufactured by Parkes of Birmingham, but retailed in Scotland.

12 Ibid., p. 121 for examples sold by Bryson of Edinburgh, but made by Wasserlein of Berlin, and by Prazmowski & Hartnack of Paris.

13 Lucio Sponza, *Italian Immigrants in Nineteenth Century Britain: Realities and Images* (Leicester University Press, Leicester, 1988), p. 61; see also pp. 80–7.

14 Alice Clark, *Working Life of Women in the Seventeenth Century* (Routledge and Kegan Paul, London, 1982), pp. 150–235.

15 Ivy Pinchbeck, *Women Workers and the Industrial Revolution 1750–1850* (Virago Press, London, 1981), pp. 282–305, esp. p. 284: 'It is only when we come to the skilled artisan and trading classes, however, that we find women still taking a share in their husbands' concerns as a matter of course, and in almost every trade innumerable instances can be cited of widows and single women in business.'

16 Ibid., p. 285.

17 Sally Alexander, 'Women's work in nineteenth-century London; a study of the years 1820–50' in Juliet Mitchell and Ann Oakley (eds), *The Rights and Wrongs of Women* (Penguin, Harmondsworth, 1976), pp. 59–111.

18 Geoffrey Crossick, 'The petite bourgeoisie in nineteenth-century Britain: the urban and liberal case' in Crossick and Haupt (eds), *Shopkeepers and Master Artisans*, pp. 62–94.

19 By 1900 Sir Howard Grubb manufactured telescopes in Dublin which were sold to an international clientele: J. Burnett, 'Thomas and Howard Grubb' in A. D. Morrison-Low and J. Burnett, *'Vulgar and Mechanick': The Trade in Scientific Instruments in Ireland 1650–1921* (Royal Dublin Society and National Museums of Scotland, Dublin and Edinburgh, 1989), pp. 89–117. By the same date, the Cambridge Instrument Company was able to market specialist instruments world-wide: M. J. G. Cattermole and A. F. Wolfe, *Horace Darwin's Shop: The Cambridge Instrument Company* (Adam Hilger, Bristol, 1987).

20 Hackman, 'The nineteenth-century trade', pp. 78–9. Charles Booth, 'Occupations of the people of the United Kingdom, 1801–1881', *Journal of the Statistical Society*, 49 (1886), pp. 314–444, esp. p. 358 for 'Watches, instruments and toys' (see appendix); and W. A. Armstrong, 'The use of

information about occupation' in E. A. Wrigley (ed.), *Nineteenth-Century Society. Essays in the Use of Quantitative Methods for the Study of Social Data* (Cambridge University Press, Cambridge, 1972), pp. 226–310, esp. p. 268 for 'Watches, instruments and toys' (see appendix).

21 Armstrong, 'The use of information'.

22 M. A. Crawforth, 'Instrument makers in the London Guilds', *Annals of Science*, 44 (1987), p. 331.

23 A discussion of their history, methods of compilation, and the problems of using them as a historical source, is to be found in Jane E. Norton, *Guide to the National and Provincial Directories of England and Wales excluding London, published before 1856* (Royal Historical Society, London, 1950); a recent survey of Irish directories is by Rosemary ffolliott and Donal F. Begley, 'Guide to Irish Directories' in Donal F. Begley (ed.), *Irish Genealogy. A Record Finder* (Heraldic Artists, Dublin, 1981), pp. 75–106. Using the Dublin directories as a starting-point, Imelda Brophy, 'Women in the workforce' in David Dickson (ed.), *The Gorgeous Mask: Dublin 1700–1850* (Trinity History Workshop, Dublin, 1987), pp. 51–63, found that most identifiable female enterprise was connected with textiles: instrument making formed too small a sample to show.

24 See Morrison-Low and Burnett, *'Vulgar and Mechanick'*. Very few specific cases of women running instrument-making or retailing firms in Ireland have been uncovered: Mrs Margaret MacHugh was recorded at 78 George's Street, Cork, between 1863 and 1883 (ibid., p. 151); Mary Cappo, widow or daughter of Joseph Cappo of Belfast, appeared in 1884 only (ibid., p. 146); Ann Voster of Cork, widowed in 1760, promptly wound up her husband Daniel's business (ibid., pp. 72, 156); while Widow Sweeny, also of Cork, kept the business going so that another male member of the family, probably her son, could take over in due course (ibid., p. 156).

25 For an overview of the Scottish trade, see Bryden, *Scottish Scientific Instrument Makers* and for instances of specific businesses, mostly from the nineteenth century, see Clarke, Morrison-Low and Simpson, *Brass & Glass*. Again, the numbers of women identified as running instrument businesses are few.

26 Jenny Wetton (of the Museum of Science and Industry, Manchester), 'Instrument making in Manchester' (forthcoming). I am grateful to Ms Wetton who has made available to me her prepublication notes about female instrument practitioners in the Manchester area. These include the daughters of John Benjamin Dancer (1812–87), Elizabeth Eleanor and Anna Maria, who took over the manufacture of microphotographs (photographs taken and viewed through a microscope, a precursor of microfilm) after their father went blind and retired in 1878. In 1896, the microphotograph negatives were sold to another business. Jacob Franks, optician and instrument-maker,

died in 1846, and his business was run for a year by his widow, Amelia. The following year she took their son Abraham into partnership, and subsequently another son, Joseph; in 1851 a third son, Henry, joined his mother and brothers in the firm, by now known as Franks & Co. Elizabeth Peduzzi (née Ward) ran her late husband's business as looking-glass and picture-frame-makers, barometer- and thermometer-makers, from his death in 1864 until her own in 1870.

27 No overall study looks at the trade in the English provinces, although E. G. R. Taylor, *The Mathematical Practitioners of Hanoverian England 1714–1840* (Cambridge University Press, Cambridge, 1966), despite its title, gives biographies of makers from all over the British Isles. For a discussion of the problems of using her work, and also the problems of using trade directories, see M. A. Crawforth, 'Makers and dates', *Bulletin of the Scientific Instrument Society*, 13 (1987), pp. 2–8. Until 1988, Michael Crawforth ran a computer data-base called 'Project SIMON', a programme to research the British scientific instrument industry from 1550 to 1850 and to record on computer a national archive of data on instrument makers and retailers, under the direction of the Senior Assistant Curator of the Museum of the History of Science, University of Oxford. Although his work was cut short by his untimely death in 1988, he provided me with a list of over 50 women he had recorded as active in the industry between 1800 and 1900 (Crawforth, private communication, October 1987). I should like to record my thanks to him and his wife and collaborator, Diana, for all their help.

28 Hackman, 'Nineteenth-century trade'; M. Dorothy George, *London Life in the Eighteenth Century* (Penguin, Harmondsworth, 1985), pp. 158–212; although dealing with an earlier period, gives a good background to the early nineteenth century: she discusses, for instance, the subdivision of labour in the watch trade.

29 Hayden J. Downing, 'Scientific instrument makers of Victorian London 1840–1900' (Museum of Victoria, Australia, 1984), p. 14: typescript in Guildhall Library, London. Downing used trade indexes of *Kelly's Post Office London Directory* for this period, backed by the commercial indexes, to extend the dates listed in Taylor, *Mathematical Practitioners*.

30 Downing, 'Scientific instrument makers', p. 16.

31 Women who appear at the same address following the same trade as a man with the same surname are assumed to be the widow carrying on the business: in this case, David Steel made his final appearance in *The Post-Office Annual Directory for 1802* (W. and J. Richardson, London, 1802).

32 *Kelly's Post Office London Directory*, 1838–46.

33 Ibid., 1813–27.

34 Ibid., 1820–7.

35 Downing, 'Scientific instrument makers', p. 54.

36 Taylor, *Mathematical Practitioners*, p. 407, gives her dates as 1819–22. However, she does not appear in the *Post Office Directory* for these years; but in *Kent's Original London Directory*, 1819, and ibid., 1823.

37 *Kelly's London Post Office Directory*, 1820–7.

38 Ibid., 1849.

39 Ibid., 1830–1.

40 Taylor, *Mathematical Practitioners*, pp. 447, 448, 452. I have been unable to trace the directories in which these were recorded.

41 *Kelly's Post Office London Directory*, 1843–5.

42 Downing, 'Scientific instrument makers', p. 26.

43 Ibid., p. 37.

44 *Kelly's Post Office London Directory*, 1838–42.

45 Ibid., 1841–5.

46 Downing, 'Scientific instrument makers', p. 21; Taylor, *Mathematical Practitioners*, pp. 425, 474.

47 *Kelly's Post Office London Directory*, 1831–8.

48 Downing, 'Scientific instrument makers', p. 34; Taylor, *Mathematical Practitioners*, p. 375 states that John Snart's 'successor is called Neazieh or Neariah in the directories of 1830–8, and was probably the Nehemia Snart of the Post Office Directory of 1846.'

49 Downing, 'Scientific instrument makers', p. 18.

50 Ibid., p. 25.

51 Clarke, Morrison-Low and Simpson, *Brass & Glass*, pp. 252–74.

52 Douglas J. Hallam, *The First Hundred Years. A Short History of Rabone Chesterman Limited* (Rabone Chesterman, Birmingham, 1984), pp. 17–20.

53 Michael A. Crawforth, *Weighing Coins: English Folding Gold Balances of the 18th and 19th Centuries* (Cape Horn Trading Company Ltd., London, 1979), pp. 145–6; Crawforth, 'Makers and Dates', p. 6. See also D. F. Crawforth, 'London scalemakers: the transmission of scalemaking knowledge from 1636–1800', *Equilibrium* (Spring 1980), pp. 207–15; (Summer 1980), pp. 239–48, 242–3 discussed women actively involved in the scale-making business, including Mary De Grave, whose husband, Charles, died in 1799, and mentions that the Blacksmiths' Company Rolls indicate 17 widows who took over businesses, were legally responsible for apprentices, or had trade labels in their own names.

54 Taylor, *Mathematical Practitioners*, pp. 304, 355: she dates W. & T. M. Bardin *c.*1780–1800, and T. M. Bardin 1806–27, but again, exact sources for her dates are not given. *The Post Office Annual Directory*, 1800–20 lists 'Bardin & Sons, globemakers'. John Millburn mentions the will of Thomas Marriott Bardin 'a well-known globe-maker of Salisbury Square who died in 1819': John R. Millburn, 'British archives for the history of instruments', *Bulletin of the Scientific Instrument Society*, 21 (1989), p. 4.

55 Downing, 'Scientific instrument makers', p. 37; Taylor, *Mathematical Practitioners*, pp. 350–1.

56 Downing, 'Scientific instrument makers', p. 35.

57 Ibid., p. 38.

58 Leonore Davidoff and Catherine Hall, *Family Fortunes: Men and Women of the English Middle Class, 1780–1850* (Hutchinson, London, 1987), esp. part 2: 'Economic Structure and Opportunity', pp. 193–315.

59 Geoffrey Crossick and Heinz-Gerhard Haupt, 'Shopkeepers, master artisans and the historian: the petite bourgeoisie in comparative focus' in Crossick and Haupt (eds), *Shopkeepers and Master Artisans*, p. 4.

60 British patent 1259, 27 June 1780, 'Constructing hydrometers with sliding-rules, to ascertain the strength of spiritous liquors, malt worts, and wash for fermentation.'

61 Francis G. H. Tate and George H. Gabb, *Alcoholometry. An Account of the British Method of Alcoholic Strength Determination* (His Majesty's Stationery Office, London, 1930), pp. 7–8.

62 John Dicas was described in Liverpool trade directories as a merchant in 1774 and 1777; a liquor-merchant in 1781; a brandy-merchant in 1787; mathematical instrument and navigation shop in 1790 and patent hydrometer and mathematical instrument maker in 1796. He died in 1797 aged fifty-six, and was succeeded by his daughter Mary Dicas. She was recorded in the Liverpool directories as patent hydrometer and mathematical instrument maker from 1800 to 1805. In 1807, the directory entry was for Dicas & Arstall, patent hydrometer makers. George Arstall was subsequently recorded as a scale-, or beam-maker, and maker of Dicas hydrometers between 1808 and 1816. The 1814 edition of *Directions for using the Patent Saccharometer* (M. Galway and Co., Liverpool, 1814) states on the title-page that it was 'invented by the late John Dicas, And for upwards of sixteen years, made only by his Daughter and Successor, Mary Arstall, late Mary Dicas, Mathematical Instrument Maker, Liverpool: the only proprietor of the patent, who, for some time previous to the decease of her Father, assisted in making the above instruments. The Manufactory of the Patent Saccharometers, Hydrometers, and Lactometers, was for some time carried on by M. Arstall under the Firm of Dicas & Co.' The text indicates that by this time instruments were stamped 'Arstall late Dicas patentee'; one copy seen has a MS note 'now made by Ann Dicas, successor to M. Arstall'. Ann Dicas appeared in the Liverpool directories as 'patent hydrometer manufactor [*sic*]' in 1817, and patent hydrometer maker from 1818 to 1821; she married Benjamin Gammage in May 1821, who was recorded in the directories between 1823 and 1871 variously as manufacturer of Dicas's patent hydrometers and saccharometers, stationer and toy-dealer; hydrometer maker (Dicas patent) and bookkeeper; and finally, as 'gentleman'. I am

grateful to D. J. Bryden for all this information. The International Genealogical Index reveals that George Arstall and Mary Dicas had four children, including Frederick Dicas Arstall who himself had offspring (fiche B0169, Aug. 1978); and that Benjamin Gammage had four daughters by his wife Ann (fiche B0224, Aug. 1978).

63 Ibid., Taylor, *Mathematical Practitioners*, p. 418 presumes she is a widow.

64 T. G. Smith, 'Bartholomew Sikes and the selection of his hydrometer for official use' (unpublished paper, H.M. Customs and Excise, Library, King's Beam House, Mark Lane, London, 1977).

65 R. D. Connor, *The Weights and Measures of England* (Her Majesty's Stationery Office, London, 1988), pp. 257–8.

66 Public Record Office, DSIR 26/92, letter from Executors of R. B. Bate to the Commissioners of Her Majesty's Revenue of Excise, 6 Jan. 1848. I am grateful to Dr Anita McConnell for this reference.

67 PRO, DSIR 26/90, Memorial of Edward Hall and Edward Jenkin, trading under the firm of Messrs Dring, Fage and Company, to the Honorable Commissioners of Her Majesty's Board of Inland Revenue, 1 May 1850.

68 'The firm of Messrs Dring & Fage has been long established, and is well known for its good manufacture of mathematical and Gauging instruments; Mr Dring, however, resigned about the time Sykes' [*sic*] Hydrometer was introduced as the legal instrument, and we believe the business was carried on afterwards by Mr Fage to the time of his death, and, subsequently, by his Granddaughter, and her Husband, and from the period of her Husband's death, by the Memorialists': PRO, Memorial of Edward Hall and Edward Jenkin. Further undated information gives the firm's succession thus: 'Established in 1745 by John Clarke carried on by his Son Richard Clarke and then by His Brother-in-law John Dring who took as Partner William Fage[.] William Fage's wife Mary Fage, left the business to her daughter Susan Fage and on her death to William George, the eldest Son of her daughter Mary[.] 1839 William George left the business to Miss Julia Holmes who took into partnership Mr Edward Hall who left the business to his daughter who married Alfred Jones who left the business to Edward Jones who left it to Mrs Jones.' Science Museum, London, Dring & Fage Archive 5/4.

69 See, for example, Davidoff and Hall, *Family Fortunes*, pp. 321–9; Lawrence Stone, *The Family, Sex and Marriage in England 1500–1800* abridged edn (Penguin, Harmondsworth, 1985); and the chapter 'Marriage' in F. M. L. Thompson, *The Rise of the Respectable Society. A Social History of Victorian Britain 1830–1900* (Fontana, London, 1988), pp. 85–113.

70 Olivia Brown, 'The instrument-making trade' in Roy Porter et al., *Science and Profit in 18th-Century London* (Whipple Museum of the History of Science, Cambridge, 1985), p. 35; Hugh Barty-King, *Eyes Right. The Story of Dollond & Aitchison Opticians 1750–1985* (Quiller Press, London,

1986), pp. 42–3. The International Genealogical Index (fiche C0179, Aug. 1978) records that Sarah Dollond married Jesse Ramsden, 16 August 1766, at St Martin-in-the-Fields, Westminster.

71 Taylor, *Mathematical Practitioners*, pp. 348, 426; G. L'E. Turner, 'Powell & Lealand: trade mark of perfection', *Proceedings of the Royal Microscopical Society*, 1 (1966), pp. 173–83; R. H. Nuttall, 'Powell and Lealand, 1900–1925', *Microscopy*, 32 (1972), pp. 158–61; David Young, 'The Firm of Powell and Lealand', *Bulletin of the Scientific Instrument Society*, 9 (1986), p. 8. The International Genealogical Index (fiche C0268, Aug. 1978) records that Elizabeth Lealand married Hugh Powell on 3 June 1824 in St Pancras Old Church.

72 Nuttall, *Microscopes from the Frank Collection*, p. 46; Mari Williams, 'Technical innovation: examples from the scientific instrument industry' in Jonathan Liebenau (ed.), *The Challenge of New Technology. Innovation in British Business since 1850* (Gower Publishing Company, Aldershot, 1988), p. 14. According to the International Genealogical Index (fiche C0169, Aug. 1978) John Henry Dallmeyer married Hannah Ross, 7 December 1854, at St Andrew's, Holborn.

73 Clarke, Morrison-Low and Simpson, *Brass & Glass*, p. 281.

74 Nicholas Goodison, *English Barometers 1680–1860* (Antique Collectors' Club, Woodbridge, 1977), pp. 309, 354–5. Also Wetton, 'Instrument making in Manchester'.

75 Crawforth, 'Instrument makers in the London guilds', p. 331.

76 Anon., 'Where Lord Kelvin's instruments are made', *The Ludgate*, 7 (1898), pp. 148–54.

77 Duncan Bythell, *The Sweated Trades. Outwork in Nineteenth-Century Britain* (Batsford, London, 1978), p. 142.

78 Downing, 'Scientific instrument makers', p. 37. Taylor, *Mathematical Practitioners*, pp. 461–2, states that George Taylor was active between 1833 and 1845, and Mrs Taylor from 1833 to 1859.

79 British patent 6582, 27 March 1834, 'Instruments for measuring angles and distances, applicable to nautical and other purposes.'

80 Taylor, *Mathematical Practitioners*, pp. 461–2; K. R. Alger, *Mrs Janet Taylor 'Authoress and Instructress in Navigation and Nautical Academy' (1804–1870)* (LLRS Publications, London, Fawcett Library Papers No 6, 1982). *Official Descriptive and Illustrated Catalogue of the Great Exhibition 1851* (Spicer Brothers, London, 1851), vol. 1, p. 339: 'a bronze binnacle, with compass, designed from the water-lily'; and ibid., p. 449: 'sextant for measuring angular distances between the heavenly bodies.' The juries felt that this latter instrument was 'intended rather for show than use': *Exhibition of the Works of Industry of All Nations, 1851. Reports of the Juries* (William Clowes and Sons, London, 1852), p. 252. There is no entry for Mrs Taylor in *The Illustrated*

Catalogue of the International Exhibition of 1862. The Industrial Department, British Division, 2 vols, (Printed for Her Majesty's Commissioners, London, 1862) – presumably she missed the printer's deadline – but her efforts were commented on by the juries in *International Exhibition 1862. Reports by the Juries* (William Clowes and Sons, London, 1863) Class XII, p. 8: a liquid compass; and ibid., Class XIII, pp. 14–15, an improved sextant. Mrs Taylor won no medals nor commendations. I am grateful to Dr Anita McConnell for these last three references.

81 The National Maritime Museum, Greenwich, has a mariner's compass (ACO.17) two octants (S.116 and S.203) and a sextant (S.81); the Science Museum, London has a sympiezometer, a form of air barometer (1908–81); and Whitby Museum has an octant.

82 For instance, 'Mme veuve Gambey' of Paris: 'Après la mort de Gambey, en 1847, on a continué à rechercher partout les instruments construits dans la maison qu'il avait fondée. Avec les machines-outils créés par Gambey, avec ses bons procédés de travail, avec les ouvriers qu'il avait formés, il a été possible, sous l'administration active et intelligente de sa veuve, et la direction de son frère, de faire d'excellents instruments de tout genre. On trouve, en effet, que les instruments exposés par Mme veuve Gambey, et si justement admirés, sont, de tout point, dignes de ceux qui sortaient des mains de notre grand artiste': *Exposition Universelle de 1855. Rapports de Jury Mixte International* (Imprimerie Impériale, Paris, 1856), p. 400. H. Lelièvre of Paris, a firm which produced French and English linear measures, and precision instruments for lineal drawing were described as having 'obtained two bronze medals . . . at the Universal Exhibition of 1867 . . . The house is supplied by several hydraulic and steam manufactories at Sommedieue, near Verdun (Meuse), employing a vast number of artificiers of both sexes': *Exposition Maritime Internationale du Havre 1868. Rapports du Jury International, et Catalogue Officiel des Exposants Récompensés* (J. M. Johnson and Sons, London, 1868), p. 55. In 1906, Elliott Brothers Ltd. of London, founded in 1800 as drawing instrument makers, were described as employing 'some 420 workpeople, including about 20 girls . . . There is a light and airy drawing office, in which six or eight draughtsmen, including a couple of girl tracers, are employed': *Milan International Exhibition 1906 Catalogue of the British Section* (London, 1906), p. 76. I am grateful to Dr Anita McConnell for all these references. Incidentally, the family concern of Elliott Brothers was run by the widow of Frederick Elliott from 1873 until her death in 1881, although for part of that time she was in partnership with W. O. Smith, who continued it with his father: Trevor Woodman and John Kinnear, 'One hundred and eighty years of instrument making. Some historical aspects of Elliott Brothers (London) Ltd. and Fisher Controls Ltd.', *Radio and Electronic Engineer*, 52 (1982), pp. 165–6. I am grateful to Dr Mari Williams for this reference.

Electronic Engineer, 52 (1982), pp. 165–6. I am grateful to Dr Mari Williams for this reference.

83 Ernest Laugier, 'Discours de M. Laugier prononcé aux funerailles de M. [Johann] Brunner, artiste du Bureau des Longitudes, le 1er décembre [1862]', *Annuaire du Bureau des Longitudes* (1863), p. 387. I am grateful to Dr Anita McConnell for this reference.

84 Clarke, Morrison-Low and Simpson, *Brass & Glass*, pp. 165–9.

85 Ibid., pp. 123–9.

86 British patent 2826, 9 March 1805. See also, I. Lovi, *Directions for using the Patent Aerometric Beads* (Printed for the Author, Edinburgh, 1813); D. J. Bryden, *Scottish Scientific Instrument-Makers*, pp. 35–6. There is a large set of Mrs Lovi's beads in the Royal Museum of Scotland (inventory T1962.115), and another in the Castle Museum, York; a third smaller set is in the Science Museum, London (inventory 1948–23), and a fourth has been noted in a private collection.

87 For instance, Goodison, *English Barometers*, mentions eight nineteenth-century female 'barometer makers': Mrs Ann Camotta, Halifax, *fl.* 1860, widow of Richard, *fl.* 1830–41; Ann Coleman, Clerkenwell, *fl.* 1832–3, widow of Charles, *fl.* 1823–9; Harriet Graham, London, *fl.* 1832–3; Mrs E. Martinelli, Lambeth, *fl.* 1852–3, widow of Alfred, *fl.* 1839–d. 1851; Margaret Pensa, Hatton Garden, *fl.* 1840–8, widow of John, *fl.* 1830–9; Jane Pizzi, Leather Lane, London, *fl.* 1840–5, widow of Valentine, *fl.* 1835–d. 1840, and partner with Henry Negretti, 1840–5; Mrs F. Tyler, Hatton Garden, *fl.* 1857–60, widow of James, *fl.* 1844–56; Elizabeth Wisker, York, 1822–7, widow of John, *fl.* 1804–d. 1822, succeeded with son Matthias to the business.

PART II

Gender Representation in Science

4

The Private Life of Plants: Sexual Politics in Carl Linnaeus and Erasmus Darwin

Londa Schiebinger

In her wane beauty, Ninon won
With fatal smiles her gay unconscious son.
Clasp'd in his arms she own'd a mother's name,
'Desist, rash youth! restrain your impious flame,
'First on that bed your infant form was press'd,
'Born by my throes, and nurtured at my breast.'
Back as from death he sprang, with wild amaze
Fierce on the fair he fix'd his ardent gaze;
Dropped on one knee, his frantic arms outspread,
And stole a guilty glance toward the bed;
Then breath'd from quivering lips a whisper'd vow
And bent on heaven his pale repentant brow;
'Thus, thus!' he cried, and plung'd the furious dart,
And life and love gush'd mingled from his heart.

Erasmus Darwin, The Loves of the Plants (1789)

Hermaphroditic plants 'castrated' by unnatural mothers. Trees and shrubs clothed in 'wedding gowns.' Flowers spread as 'nuptial beds' for a verdant groom and his cherished bride. Are these the memoirs of an eighteenth-century Academy of Science, or tales from the boudoir?

These are, in fact, some of the categories developed by Carl Linnaeus, the father of modern botany, and his disciples in their attempts to understand the sexuality of plants. Plant sexuality was not incidental to but indeed lay at the heart of the eighteenth-century revolution in the study of the plant kingdom. From the Middle Ages through the

PLATE 4 'Cupid Inspiring the Plants with Love.' Cupid, the god of love, aims his arrow at the Bird of Paradise (*Strelitzia reginae*) or the 'Queen Flower' as it was called in honor of Queen Charlotte, to whom the book is dedicated.

(Illustration from Robert Thornton's *Temple of Flora*, London, 1805).

Renaissance there had been one primary reason for studying botany: plants were used as medicines. In the eighteenth century, however, focus shifted from the medicinal virtues of plants to finding abstract and universal methods of classification. Paradoxically, this 'scientization' of botany coincided with an ardent 'sexualization' of plants.

Why did the study of plant sexuality become a priority of the botanical sciences in the eighteenth century? More importantly, why was it sexuality that became the key to classification? Botanists praised Linnaeus for creating a new language for botany. Linnaeus's nomenclature was indeed precise and convenient; Rousseau judged its value to botanists as great as the creation of algebra for mathematicians.[1] The notion that plants are sexual did not arise uncontested, however. Great debates erupted in the late eighteenth and early nineteenth centuries over the scientific and moral propriety of what Erasmus Darwin (grandfather of Charles) called 'the loves of the plants.' Yet, it is important to point out that these self-appointed censors missed the sexualists' most glaring impropriety – that of reading the laws of nature through the lens of social relations. Indeed, as we shall see, the fundamental divisions Linnaeus devised as the basis of his new botanical taxonomy recapitulated the sexual hierarchy of Western Europe.

In this essay, I distinguish two levels in the sexual politics of early modern botany – the *implicit* use of gender to structure botanical taxonomy and the *explicit* use of human sexual metaphors to introduce notions of plant reproduction into botanical literature. I sketch first the most important aspect of sexual politics in early modern botany – the way that unarticulated notions of gender structured Linnaean taxonomy. In the uproar that surrounded the introduction of notions of sexuality into botany, no one noticed that Linnaeus's taxonomy – built as it was on sexual difference – imported into botany traditional notions about sexual hierarchy. Secondly, I explore the origins and implications of the vivid sexual language employed by Carl Linnaeus and Erasmus Darwin. Investigating the specific circumstances surrounding the reception of each of these botanists in England deepens our understanding of how science interacts with changing cultural situations, such as changing attitudes toward women and female sexuality, the stability of Church and State, and the like. But, as I argue, the sexual politics of botany in the eighteenth century cut even deeper into the political landscape, having ultimately to do with the European-wide revolution in scientific views of sexual difference that took place in the upheavals leading up to the French Revolution.

THE SEARCH FOR NEW METHODS OF CLASSIFICATION

Within medieval cosmology, plant classification generally emphasized the usefulness of plants to human beings as foods and medicines. Even in the seventeenth century, botany remained closely allied with medicine; herbal texts classified plants according to their use. Plants were often arranged alphabetically, each entry including a description of its appearance and varieties, the season and place it could be collected, the parts to be used and how to preserve them. Also included would be the degrees of heat and moisture, its powers against particular ailments, doses, and methods of preparation and administration.[2] Knowledge of plants at this time was local and particular, derived from direct experience with plants in agriculture, gardening, or medicines, or knowledge based upon that experience.

In the seventeenth century, academic botanists began to break their ties with medical practitioners. New plant materials from the voyages of discovery and the new colonies flooded Europe at the same time that an emphasis on observation increased discord between ancient texts and modern knowledge. The proliferation of knowledge required new methods of classifying that knowledge. Emphasis in classification turned from medical application to more general and theoretical issues of pure taxonomy, as botanists sought simple principles for classification which would hold universally.[3]

That sexual differentiation in plants was to provide this unifying principle was not immediately apparent. Though botanists agreed that a new nomenclature was desirable, few agreed on the system. By 1799, when Robert Thornton published his popular version of the Linnaean system, he counted 52 different systems of botany; the 'system-madness,' one botanist complained, was truly 'epidemical.'[4] Botanists based their systems on different parts of the plant. In England, John Ray developed a system for establishing genera based on the flower, calyx, seed, and seed-coat; in France the great Jospeh Pitton de Tournefort defined genera principally by the characteristics of the corolla and the fruit. Sébastien Vaillant and Linnaeus were early advocates of classification by sexual differences. Others, including Albrecht von Haller, continued to argue that geography was crucial to an understanding of plant life and that development as well as appearance should be represented in a system of classification.

Despite the number and variety of systems, Linnaeus's sexual system was widely adopted after 1737 and until the first decades of the nineteenth century was generally considered the most precise and convenient system of classification. Linnaeus's system, as set out in his first major publication, *Systema naturae* (1735), was based on the difference between the male and female parts of flowers. Linnaeus divided the vegetable world (as he called it) into *classes* based on the number, relative proportions, and position of the male parts or stamens. These classes were then subdivided into some 65 *orders* based on the number, relative proportions, and position of the female parts or pistils. These were further divided into *genera* (based on the calyx, flower, and other parts of the fruit), *species* (based on the leaves or some other characteristic of the plant), and *varieties*.

One might argue that Linnaeus based his system on sexual difference because he was one of the first to recognize the biological importance of sexual reproduction in plants. For this reason, he considered the stamens and pistils 'the very essence of the flower.'[5] But the success of Linnaeus's system did not rest on the fact that it was 'natural,' that it captured true affinities between organisms. Indeed, Linnaeus readily acknowledged that it was highly artificial.[6] Though focused on reproductive organs, his system did not capture fundamental sexual functions. Rather it focused on purely morphological features (that is, the number and mode of union) – exactly those characteristics of the male and female organs *least* important for their sexual function.[7]

In view of this fact, it is striking that Linnaeus chose to highlight the sexual parts of plants at all. Furthermore, it is important to point out that Linnaeus devised his system in such a way that the number of a plant's stamens (male parts) determined the *class* to which it was assigned, while the number of its pistils (the female parts) determined its *order*. In the taxonomic tree, class stands above order. In other words, Linnaeus gave male parts priority in determining the status of the organism in the plant kingdom. There is no scientific justification for this outcome; rather, Linnaeus brought traditional notions of gender hierarchy whole-cloth into science. He read nature through the lens of social relations in such a way that the new language of botany incorporated fundamental aspects of the social world as much as those of the natural world. Though today Linnaeus's classification of groups above the rank of genus has been abandoned, his binomial system of nomenclature remains, together with many of his genera and species.

But the debate surrounding Linnaeus's notions of plant sexuality did not focus on this fundamental (mis)reading of the laws of nature. No one at the time or since then, for that matter, objected to this flaw in his system. Rather debate centered on scientific and moral questions surrounding the nature of plant sexuality and the language Linnaeus and his disciples used to describe the loves of the plants.

THE USE OF METAPHOR IN SCIENCE

Linnaeus's system focused as much on the 'nuptials' of living plants as on their sexuality. Before their lawful marriage, trees and shrubs donned 'wedding gowns.' Flower petals spread as 'bridal beds' for a verdant groom and his cherished bride, while the curtain of the *corolla* lent privacy to the amorous newly-weds. Other plants engaged in 'adulterous' relations. This desire of botanists to view plants as highly erotic creatures reached its peak in the work of Erasmus Darwin. In his poem *The Loves of the Plants* an ordinary *gloriosa superba* repulses the incestuous advances of her son, who, clasped in her arms, steals a guilty glance toward the bed of his passion and, quivering, plunges a dagger into his own heart.[8]

The question of the nature of Linnaeus's and Darwin's sexual metaphors – their social context and implications – has been a topic of debate among historians of science. Some have dismissed them as 'charming absurdities.'[9] Others have suggested that the Enlightenment was a time of freer sexuality and that attributing sexuality to plants was designed to titillate appreciative audiences.[10] Still others have taken a more prudish view, suggesting that sexuality was acceptable only if plants observed the mores of eighteenth-century society, having sexual relations only after 'lawful marriages'.[11]

The origins and implications of this imagery are complex. Its use was not bounded nationally: Vaillant, a Frenchman, used it in 1717; the Swede Linnaeus never abandoned it; Charles Bonnet, the Swiss naturalist, also embraced it; the Englishman Erasmus Darwin brought it to its peak in the 1780s and 1790s. This imagery was not, as one might imagine, merely the product of young men's minds: Vaillant and Darwin both wrote toward the end of their lives. Nor were there clear political alignments. In England, conservatives such as the Reverend Richard Polwhele associated loose sexual imagery with Jacobin free love, and Erasmus Darwin, a radical democrat and atheist, may have tried through his revelations of plant polygamy to subvert straight-laced middle-class monogamy.[12] Linnaeus, by contrast, was distinctly conservative in his social attitudes. He wanted his daughters to grow up to be hearty, strong housekeepers, not 'fashionable dolls' or blue-stockings.

Generalizations thus do not easily emerge about the meaning of this imagery. Attitudes toward sexuality – whether in plants or humans – were influenced by changing attitudes toward women and female sexuality, the stability of Church and State, and the importance attributed to sex as an explanatory device. We can best understand these shifting circumstances by comparing the reception of Linnaeus's work in England to the changing fortunes of Darwin's *The Loves of the Plants*.

The historian François Delaporte has suggested that in the eighteenth century plant sexuality had to be contained within lawful marriages, and it is true that for Linnaeus plant sexuality took place almost exclusively within the bonds of marriage. When Linnaeus introduced new terminology to describe the sexual relations of plants, he did not use the terms stamen and pistil rather *andria* and *gynia*, the Greek terms for husband and wife. The names of his classes of plants ended in 'andria' (*monandria, diandria, triandria*, and so on); his orders end in 'gynia' (*monogynia, digynia, trigynia*, and so forth). His text is filled with tender embraces of duly wedded couples:

> The flowers' leaves . . . serve as bridal beds which the Creator has so gloriously arranged, adorned with such noble bed curtains, and perfumed with so many soft scents that the bridegroom with his bride might there celebrate their nuptials with so much the greater solemnity. When now the bed is so prepared, it is time for the bridegroom to embrace his beloved bride and offer her his gifts . . . [13]

Indeed, his renowned 'Key to the Sexual System' is founded on the *nuptiae plantarum* (the marriages of plants); the plant world is divided into major groups according to the type of marriage each plant has contracted – whether, for example, they have been wed 'publicly' or 'clandestinely.' (It is interesting to note that these two types of marriage did characterize custom in much of Europe; only in 1753 did Lord Harwicke's Marriage Act do away with clandestine marriages by requiring a public proclamation of banns.)

Delaporte has overlooked the fact, however, that the majority of Linnaeus's plants do not engage in lawful marriages. Only one class of plants – Linnaeus's *monandria* – practises monogamy. Plants in other classes engage in marriages consisting of two, three, twenty, or more 'husbands' who share their marriage-bed (that is, the petals of the same flower) with one wife. Plant husbands of his 'class xxiii' – *polygamia* – live with their wives and harlots, later called concubines, in distinct marriage beds. Each of these 'marriages' signifies a particular arrangement of stamens and pistils on the flower. *Monandria* have but one stamen or husband on a hermaphroditic flower. *Diandria* have two stamens (or husbands) on a flower with one pistil, and so on.

These images of happily-wedded plants, though compelling, are superfluous; directly after each metaphor of marriage Linnaeus gives a flat-footed description of the arrangement of genitalia on the flower. Though gratuitous, the notion that plants and animals reproduce within marital relations persisted into the nineteenth century. The term 'gamete' – adopted by biologists in the 1860s to refer to a germ cell capable of fusing with another cell to form a new individual – derives from the Greek *gamein*, to marry.[14]

There is no evidence that Linnaeus employed sexual images with any intention of shaking loose eighteenth-century marriage practices. He was conservative in his religious views (all of nature celebrating the glory of its Creator) and in his views with respect to his daughters' upbringing and education. His four daughters were not allowed to learn French for fear that with the language they would adopt the liberties of French custom. When his wife placed Sophia in school, Linnaeus immediately took her out again, stopping what he considered 'nonsensical' education. He also refused Queen Louisa Ulrika's offer to receive one of his daughters at court, thinking the court environment apt to corrupt morals.[15] He did, however, allow his daughter to develop a mild interest in botany; Lisa Stina contributed a paper to the *Transactions of the Royal Academy of Science* entitled 'Remarks on a luminous appearance of the Indian Cresses' (or *Tropaeolum*, though she did not use the technical term).

It seems clear, then, that Linnaeus did not introduce his sexual imagery as an affront to social custom. He simply tended to see anything female as a wife. He considered 'Dame Nature' his other wife and true helpmeet.[16] The celebrated botanical illustrator, Mademoiselle Basseporte, who worked at the Jardin du Roi in Paris, he called his 'second wife'.[17] Linnaeus called his own wife 'my monandrian lily;' the lily signifying virginity, and *monandrian* meaning 'having only one man.'[18]

Though to us Linnaeus's images seem incongruous with his science, they were not completely out of place when he wrote in the mid-eighteenth century. A sharp line had yet to be drawn between scientific writing and poetic imagination, as Marina Benjamin has discussed above. Rhetorical flourish – ornamentation, metaphor, or allegory – was still appreciated in scientific texts and illustrations.[19] Early modern botanists employed a wide variety of images. The sixteenth-century Italian botanist, Andreas Caesalpinus, had arranged plants like armies ready for battle. Linnaeus portrayed nature as a copy of human society with its military and civil classes in his *Deliciae naturae*. Robert Thornton, Darwin's contemporary, used Christian imagery, seeing the passion flower as an embodiment of the passion of Christ; the leaves were said exactly to resemble the spear that pierced his side, the tendrils appeared to be the cords that bound his hands, and so forth.[20] Moreover, Linnaeus drew much of his sexual imagery directly from the classics.[21] Pliny, for example, had described the passion of love and mutual embrace of the date palms. Erasmus Darwin, too, employed a kind of Ovidian personification by transforming flowers and plants into men and women.[22] In this sense, the fathers of modern botany stood close to older traditions, being as much the denizens of an earlier era as the heralds of a new one.[23]

LINNAEUS'S RECEPTION IN ENGLAND AND THE BOTANIC REVERIES OF ERASMUS DARWIN

Linnaeus took England by storm in the 1750s and 1760s. His sexual system – which had come to dominate European botany by the middle of the eighteenth century – gained easy acceptance in Britain because academic natural history had been in decline since the 1720s; the classificatory advances of John Ray (one of the first to develop a system based on natural affinities) persisted but without further interest or development. During this same period, however, natural history – especially entomology, conchology, and eventually botany – became popular among the fashionable. Well-born ladies, including the Duchess of Beaufort, Lady Margaret of Portland, and Mrs Eleanor Glanville, led the way, collecting rare and exotic plants from all over the world.[24] The royal family (George III, Queen Charlotte, and his mother Augusta – all botanical enthusiasts) further enhanced the popularity of botany by serving as influential patrons and enlarging the Royal Botanic Gardens at Kew.

It was in this atmosphere, where botany was popular especially among ladies of the upper classes, that Linnaeus's sexual system gained wide acclaim. The *Gentleman's Magazine* lauded his work in 1754. Three popular translations of his prize-winning essay on the sexes of plants were published between 1777 and 1807.[25] The Linnaean Society was founded in 1788. The Botanical Society of Lichfield (under the auspices of Erasmus Darwin and with the aid of Samuel Johnson) sponsored a translation of both Linnaeus's *Systema naturae* and *Species plantarum*.

But it was also in Britain that bloody and protracted battles erupted almost immediately over the scientific and moral implications of Linnaeus's sexual system. Anti-sexualists, especially those working in Edinburgh, attacked Linnaeus's work primarily on empirical grounds. Charles Alston, professor of medicine and botany at the University of Edinburgh, argued in his address to the Philosophical Society in 1754 that there was simply not sufficient data to enable botanists to attribute a prolific role to pollen. The crucial experiment for anti-sexualists was to discover whether a female plant can produce fertile seeds without coming into contact in any way with the male; this they claimed was indeed the case.[26]

William Smellie, chief compiler of the first edition of the *Encyclopaedia Britannica*, blasted the 'alluring seductions' of the analogical reasoning upon which the sexualist hypothesis was founded and argued that it did not hold up to facts of experience. Many animals (he mentioned polyps and millipedes) reproduce *a*sexually, and if many species of animals are destitute of 'all the endearments of love,' what, he asked, should induce us to fancy that the oak or mushroom enjoy these distinguished privileges? Neither could Smellie abide the notion that plants might change their sex – that trees that had been female might suddenly assume the robust features peculiar to the male. Nor did he think that nature would leave a matter of such import as reproduction to chance. He felt certain that pollen, flying 'promiscuously' aloft, would produce universal anarchy and cover the earth with 'monstrous productions.'[27]

In addition to his empirical qualms, Smellie denounced Linnaeus for taking his analogy 'far beyond all decent limits,' claiming that Linnaeus's metaphors were so indelicate as to exceed the most 'obscene romance-writer.'[28] Smellie's sentiments were shared by other botanists. William Withering, a member along with Erasmus Darwin of the Lunar Society, advocated suppressing 'the sexual distinctions in the titles to the Classes and Orders.'[29] Some years earlier in St Petersburg, Johann Siegesbeck had also found the idea of sexuality in flowers empirically unconvincing and morally revolting. What man, he fumed, would believe that God Almighty would introduce such 'shameful whoredom' into the plant kingdom?[30] John Amman, another professor of botany in St Petersburg, objected to Linnaeus's system because the great concourse of husbands to one wife is so 'unsuitable to the laws and manners of our people.'[31]

In face of such opposition, few of the authors who popularized Linnaeus's system – John Miller, James Smith, or John Rotheram – made use of his sexual imagery. Popularization was left to Erasmus Darwin, who in 1789 with *The Loves of the Plants* brought the Linnaean sexual system to full bloom, elaborating Linnaeus's ideas in such a way that may well have shocked Linnaeus himself.

Unlike Linnaeus's, Darwin's plants did not limit sexual relations to the bonds of holy matrimony. Rather they freely expressed every imaginable form of human sexuality. The fair *Collinsonia* – sighing with sweet concern – satisfied the love of two brothers by turns. The *Meadia* (an ordinary cowslip) bowed with 'wanton air,' rolled her dark eyes, and waved her golden hair as she gratified each of her five beaux. Three youthful swains succumbed to the riper years of the *Gloriosa*.[32] For Darwin, sex was not only the purest source of happiness – 'the cordial drop in the otherwise vapid cup of life' – but the mechanism for improving and diversifying the stock of living organisms.[33]

To the modern reader, Darwin's poetry seems arcane and overdrawn. Though Anna Seward, a close friend and well-known poet, praised the *Botanic Garden* for establishing a new poetic form by adapting scientific discoveries to heroic verse, Darwin's poetry was not new, nor was it esoteric or unusual.[34] The eighteenth century abounded with didactic poems on raising hops, sugar-cane, gardening, and the like.[35] Botany, too, figured as the subject of elaborate poems; one reviewer of Darwin's *Loves of the Plants* even suggested that Darwin might be guilty of plagiarizing Monsieur de la Croix's *Connubia florum*.[36]

Darwin's impulse to write about botany in verse no doubt came from Linnaeus. But Darwin was also content to cash in on the botany craze. To James Watt and the industrialists of the Lunar Society, Darwin justified his seemingly frivolous endeavour, saying that 'the Loves of the Plants pays me well, and . . . I write for pay, not for fame.'[37] At the same time, Darwin feared the 'professional danger' of writing poetry. He first published *The Loves of the Plants* anonymously in 1789 because (as he confided to his publisher) he thought it might injure his lucrative medical practice.[38] Darwin had suggested initially that Anna Seward write the poetry, a genre then considered more appropriate to a woman, while he supply elaborate scientific notes in prose. Seward, however, declined, objecting that she did not know enough about botany and that the enterprise was not 'strictly proper for a female pen' (she does not say why).[39]

Darwin finally undertook the task himself (though he did incorporate some of Seward's verse without acknowledging it as hers). Darwin's stated purpose in writing these poems was to enlist 'imagination under the banner of science.'[40] Though Seward judged *The Loves of the Plants* appropriate for philosophers (for whom the extensive notes would provide the prime interest) and fine ladies and gentlemen (who would be drawn to the charms of poetry), Darwin intended the poem for a popular audience. This was also true of his later work. When preparing his *Economy of Vegetation*, he asked James Watt for some 'gentlemanlike facts' (that is, agreeable facts and not 'abstruse calculations, only fit for philosophers') about his steam-engine.[41]

The fact that Erasmus Darwin set science to verse was not completely out of step with eighteenth-century practice. But how was his erotic verse received? It is difficult to say exactly what Darwin's own intentions were in this regard. He could hardly have been unaware that the overtly erotic scenes would appeal to a wider public already overly fond of romances. Yet, in his *Commonplace Book*, Darwin expressed his belief that Linnaeus's sexual terminology could be translated without evoking indecent ideas.[42] Anna Seward also defended Darwin against charges of indelicacy: 'If the *Botanic Garden* is to be judged immodest,' she wrote, 'the impurity is in the imagination of the reader, not on the pages of the poet.' For her, the sexual nature of plants was simply a fact of science. In fact, Darwin's poetry was criticized at the time for lacking 'sensation'; while it delighted the imagination, it was said not to excite deeper emotions.[43]

But this is not the whole story. Darwin was not the conservative that Linnaeus was. He was a founding member of the Lunar Society of Birmingham, where about half of the men involved were Dissenters, and he was himself an atheist. Members, such as Joseph Priestley and Josiah Wedgwood, fostered innovation in industry and advocated liberty, equality, and leadership for the middle classes in politics. (The working classes were to supply an orderly and well-disciplined work-force for the new factory system.)[44] Darwin himself supported religious freedom, freedom of the press, the abolition of the slave-trade, and, like most of the members of the Lunar Society, welcomed the French Revolution.[45] Darwin also chose as his publisher Joseph Johnson, who was to become well known for disseminating radical literature, including the works of Joseph Priestley, Thomas Paine, William Godwin, and Mary Wollstonecraft, during the French Revolution.

Remarkably, neither Darwin's style nor his politics disturbed the public when his *Loves of the Plants* first appeared in 1789. The 200-page poem was greeted with enthusiasm and favorably reviewed in both the Whig *Monthly Review* and the Tory *Critical Review*.[46] Similarly, his *Botanic Garden* (1791) and *Zoonomia* (a medical treatise and early systematic statement of the theory of evolution of 1794) met with success, and for a time Darwin was one of the most widely-read poets in England.

This initial indifference to Darwin's unorthodox scientific, religious, and political opinions has been explained as a result of the social stability England had enjoyed since the 1750s.[47] In this atmosphere mild expression of unorthodox opinion by men of the gentry and professional classes could safely be tolerated. As Roy Porter has argued, the Enlightenment had also ushered in more tolerant views of human sexuality.[48] Sex was no longer seen as a sin or vice, but as part of the economy of nature – a natural impulse that should find free expression. Free love was not only discussed among elites, it was practised: pornographic journals began appearing from the 1770s; erotic novels proliferated; men of substance walked in public with their mistresses; and bastards grew up as accepted members of the family (though without inheriting the family name or property). Fanny Hill even spoke of the penis as a 'sensitive plant.' Sexuality expressed within the bounds of upper-class sensibility and decorum could be tolerated because it did not pose a serious threat to social order.[49]

The French Revolution shattered this calm. Members of the Lunar Society were violently attacked in the Birmingham Riots of 1791 for their open expression of support for the Revolution. Priestley's and Withering's houses were sacked. Thereafter the society was disbanded and members sought to distance themselves from politics.[50] As Church and State increasingly felt endangered, the official reaction to Darwin's work changed. The year 1798 saw the publication of *The Loves of the Triangles* (written in mock Darwinian verse) in the *Anti-Jacobin*, a semi-official weekly, controlled by George Canning, Foreign Under-Secretary in Pitt's government. In this poem Darwin was attacked for being, among other things, an advocate of free love – a tendency that conservatives saw as threatening to undermine English society in the same way that it had French society.[51] Darwin, under the cover of poetic license, may well have been advocating the free love that he himself practised after the death of his first wife. Darwin had two illegitimate daughters – Susan and Mary Parker – whom he and his second wife raised on equal terms with their other children.

In his *Unsex'd Females*, the Reverend Richard Polwhele, Bishop of Exeter, asserted that the open teaching of the sexual system in botany encouraged unauthorized sexual unions. For him, democratic tendencies, liberated and irreligious women, and free love were all aspects of the danger facing England. Polwhele attacked that Amazonian band of 'female Quixotes of the new philosophy' for adopting the sentiments and manners of republican France, and singled out Mary Wollstonecraft as the prophetess of the movement. In a striking passage, Polwhele exploited the full potential of botanical allegory, which he so despised, in order to paint for the reader a vivid picture of Wollstonecraft's 'disgraceful' life:

> But hark! lascivious murmurs melt around;
> And pleasure trembles in each dying sound.
> A myrtle bower, in fairest bloom array'd
> To laughing Venus streams the silver shade . . .
> Bath'd in new bliss, the Fair-one [Wollstonecraft] greets the bower
> And ravishes a flame from every flower;
> Low at her feet inhales the master's sighs,
> and darts voluptuous poison from her eyes.
> Yet, while each heart pulses, in the Paphian grove,
> Beats quick to Imlay and licentious love.[52]

Polwhele is wilfully ambiguous in the poem; the reader is left uncertain whether Wollstonecraft actually becomes one of Darwin's 'adulterous' plants or if her libertine relations simply take place in the heaving atmosphere of the floral bower.

In any case, the message is clear: association with plants leads to licentious love. In his notes Polwhele explains how Wollstonecraft's liaisons in England with Henry Fuseli, the well-known painter who provided several illustrations for Darwin's *Botanic Garden*, and in France with the American writer Gilbert Imlay, led her to attempt suicide from which she was rescued by Godwin only to die soon thereafter in childbirth. This, in Polwhele's view, was a just end to a dissolute life. Our botanizing girls, he wrote, are worthy disciples of Miss Wollstonecraft. These 'unsex'd females,' sworn enemies to blushes, throw aside their modesty – that most brilliant ornament of their sex.[53]

Polwhele considered Erasmus Darwin's *Botanic Garden*, where 'lustful boys anatomize a plant,' a prime source of moral decay.[54] Though Darwin was a democrat and materialist, it should be pointed out that his radicalism with respect to women was measured. His *Plan for the Conduct of Female Education*, written for the school set up by his illegitimate daughters at Ashbourne, was in step with the new middle-class prescriptions of sexual complementarity, which advocated distinct roles for men and women in society.[55] Like Rousseau, Europe's leading advocate of sexual complementarity, and Madam Genlis, whose works he recommended to his readers, Darwin held that the female character was to possess mild and retiring virtues rather than bold and dazzling displays. Darwin emphasized that female learning and refinements were not to lead to professional employ, but were to be directed toward enhancing the home life of prospective husbands. Darwin's one stroke of innovation was his insistence on physical education for girls.[56] For the most part, however, his girls were to be schooled to be good wives – cheerful, deeply moral, and respectful of religion.

Though the Polwheles of the world taught that Linnaeus's botanical system was inherently dangerous, it was considered a threat to the social order only when combined (as in the case of Erasmus Darwin) with radical politics. Consider the example of Robert Thornton, a contemporary of Darwin, whose illustration of 'Cupid Inspiring the Plants with Love' I have had reproduced at the beginning of this essay (see plate 4). Characterizing himself as a patriot and 'lover of social order', Thornton published his *New Illustration of the Sexual System of Carolus Linnaeus* in 1799 as a challenge to French ascendancy in the fine arts.[57] Though Thornton made ample use of Darwin's poetry, his project enjoyed royal patronage. It was dedicated to Queen Charlotte, and many of the plates were done by the king's own engravers. Thornton's book sold well until 1808, when illness and old age rather than political reaction brought an end to royal enthusiasm for botany, and popular tastes began turning toward the new science of chemistry.

THE POLITICS OF SEXUAL DIFFERENCE

Much attention has been focused on the overt sexual imagery used by Linnaeus and Darwin. These tales are indeed intriguing, but they divert attention from the heart of the story – broader political trends that induced botanists to set as a priority the investigation of sexual difference in plants in the eighteenth century.

It is significant that botanists did not recognize plant reproduction as sexual until well into the eighteenth century. Certainly, as Linnaeus himself points out, the earliest observers of nature (Aristotle and, especially important for the study of plants, Theophrastus) were not ignorant of the sexes of plants.[58] Though the Greeks were aware that plants reproduce sexually, it was not the focus of their attention. As late as the Renaissance, botanists gave names to what we today call the sexual parts of flowers that were not associated in any way with their function. The male organ was called the *stamen* a latin word denoting the warp thread of a fabric. The female organ was called *pistil*, a term suggesting the resemblance of those flower parts to a pestle.[59] Renaissance herbalists simply did not recognize or were not interested in the sexual nature of plants. Why, then, did the preoccupation with sexual difference – a project that would carry forward into the nineteenth century – emerge in the eighteenth century?

Looking first within the scientific community, it is clear that the recognition of sexual reproduction in plants was pushed by developments in zoology. This was the period of keen interest in theories of generation with debates raging between preformationists and epigenesists, ovists and homunculists. Botanists, committed to finding uniform laws of nature, increasingly drew analogies between plant and animal life, and thus transferred their interest in animal sexuality to plants.

One might also conjecture that agriculture and the need for increased crop yields led to an increased interest in plant reproduction. It seems likely that the perceived need for population growth – which had sparked interest in the science of human generation (especially in France) – also fuelled botanists' interests.

Yet there is a broader context worth considering. It is no accident, I propose, that the first studies of sexuality in plants arose in the eighteenth century and coincided with a keen interest in exact differentiation of sexual character in animals and humans. Both botany and anatomy were branches of medicine as taught at the early modern university. But more than that, both were subject to the imperative to find and analyze sex (and gender) differences that dominated eighteenth-century scientific communities. This was, after all, the period of the scientific revolution in views of sexual difference.[60] Three aspects to this revolution are worth considering. First, this revolution brought about a thoroughgoing reform of definitions of sexuality. Anatomists articulated a new vision for the origins and character of sexual differences, the relation of sex and gender, and the presence of sexuality in the body. Employing new techniques, they weighed and measured every imaginable difference between the bones, hair, mouths, eyes, voices, blood vessels, sweat, and brains of men and women.

Secondly, it is important to understand that this revolution emerged as part and parcel of the political and social upheaval leading up to the French Revolution. The question for Enlightenment thinkers was: what to do with women? The sexual freedoms enjoyed by upper-class women (which, as we have seen, encouraged popular discussion of plant sexuality) and the potential freedoms demanded by women of the third estate threatened men of the emerging middle classes. If women were not to enter the new republic on an equal footing with men (and they were not), how was this to be reconciled with the Enlightenment axiom that all men are by nature equal? Increasingly the claim of women to equality was taken to be a matter not of ethics but of anatomy, as Ornella Moscucci's chapter demonstrates, and the question of the exact nature of sexual difference became a priority of the medical sciences.

Thus enlisted, anatomists offered an ideological resolution to the woman question which forms a third aspect of the revolution in sexual difference – a solution known as the theory of sexual complementarity. This theory, which grounded the ideology of separate spheres throughout much of the nineteenth century, taught that man and woman are not physical and moral equals, but complementary opposites. The theory of sexual complementarity fits neatly into dominant strands of democratic thought by making inequalities seem natural, while satisfying the needs of European society for a continued sexual division of labour.

The fact, then, that Linnaeus's system gained such currency in the eighteenth century can best be traced to the upheavals surrounding the nature and definition of sexuality and sexual roles in that century. Linnaeus focused on sexuality as his principal taxonomic division because he saw the sexual organs as the most important organs of the plant. And I would argue that he saw plants in this way because he viewed them through an eighteenth-century lens. The sexual revolution of the Enlightenment had a deep impact on the rise of botanical sciences. Not only were sexual images prominent in the language of botanists, but botanical taxonomy recapitulated prominent aspects of European sexual hierarchy.

NOTES

1 Linnaeus devised a system of binary nomenclature that was universally adopted for both plants and animals and still serves today as the backbone of biological classification. In 1905 the *International Code of Botanical Nomenclature* designated Linnaeus's *Species plantarum* of 1753 the starting-point for botanical nomenclature. See Frans A. Stafleu, *Linnaeus and the Linnaeans: The Spreading of Their Ideas in Systematic Botany, 1735–1789* (A. Oosthoek's Uitgeversmaatschappy, NV, Utrecht, 1971), p. 110.

2 See Christoph Trew, *Vermehrtes und verbessertes Blackwellisches Kräuter-Buch* (Nuremberg, 1750), preface, and Karen Reeds, 'Botany in Medieval and Renaissance Universities' (Ph.D. dissertation, Harvard University, 1975). Agnes Arber has pointed out, of course, that from the time of Aristotle plants had been studied from two diverse viewpoints: as a branch of natural philosophy and as a by-product of medicine or agriculture; see her *Herbals: Their Origin and Evolution, a Chapter in the History of Botany 1470–1670*, (3rd edn, Cambridge University Press, Cambridge, 1986), p. xxv and chapter 1.

3 Historians of botany traditionally emphasize the break between herbalism and modern botany. This change, it should be noted, was gradual, and many of those today recognized as the founders of modern botany continued to write about the medicinal qualities of plants. Linnaeus, for example, thought members of natural plant groups displayed similar medicinal properties.

4 Thomas Martyn, cited in David E. Allen, *The Naturalist in Britain: A Social History* (1976; Penguin, Harmondsworth, 1978), p. 39.

5 Carl Linnaeus, *Systema naturae* (1735) ed. M. S. J. Engel-Ledeboer and H. Engel (B. de Graaf, Nieuwkoop, 1964), 'Observationes in regnum vegetabile,' no. 7.

6 Linnaeus's ultimate goal was to devise a natural system, but until that time an artificial system – that could group the plants in an orderly fashion – was required: ibid., no. 12. See also Linnaeus to Haller of 3 April 1737, in *The Correspondence of Linnaeus*, ed. James E. Smith (2 vols, London, 1821), vol. 2, p. 229. Linnaeus's system was artificial but it was neither arbitrary nor

heterodox, see James Larson's discussion in his *Reason and Experience: The Representation of Natural Order in the Work of Carl von Linné* (University of California Press, Berkeley and Los Angeles, 1971), pp. 61–2.

7 See Julius von Sachs, *Geschichte der Botanik vom XVI. Jahrhundert bis 1860* (Munich, 1876), pp. 82–3.

8 Carl Linnaeus, *Praeludia sponsaliorum plantarum* (Uppsala, 1729), section 16. I thank Vigdis Eriksen for her help in translating this essay from the Swedish. Erasmus Darwin, *The Loves of the Plants* (1789) in *Botanic Gardens* (Garland Publishing, New York, 1978), vol. 2, pp. 14–15.

We often think of the use of metaphor and extra-scientific imagery as a thing of the past, but it still pervades science. Mary Willson, plant ecologist and sociobiologist at the University of Illinois, has recently written that her scientific language sometimes appears anthropomorphic. Willson's plants do not engage in the tender marriages characteristic of Linnaeus's or Erasmus Darwin's. Being more up-to-date, they channel their energies into their investment portfolios. Willson's plants judge their 'reproductive success' ('RS,' in the sociobiological literature), measured against the cost of their mating investment (MI) and parental investment (PI). Willson puts it this way: 'The intensity of sexual selection can be indexed by the disparity (as a ratio) of net RS (reproductive success) of breeding males and females, where net RS is measured as benefit to the parents (in present offspring) minus the cost (in reduced members or quality of future offspring).' Thus, Willson's plants have 'tactics,' make 'choices,' engage in 'conflict,' etc. Mary F. Willson and Nancy Burley, *Mate Choices in Plants* (Princeton University Press, Princeton, 1983), pp. 3–4.

9 See, for example, William T. Stern, cited in Wilfrid Blunt, *The Compleat Naturalist: A Life of Linnaeus* (William Collins Sons, London, 1971), p. 244.

10 John Farley, *Gametes and Spires: Ideas about Sexual Reproduction, 1750–1914* (The Johns Hopkins University Press, Baltimore, 1982).

11 See François Delaporte, *Nature's Second Kingdom: Explorations of Vegetality in the Eighteenth Century*, tr. Arthur Goldhammer (1979; The MIT Press, Cambridge, Mass., 1982), pp. 143–4.

12 Desmond King-Hele, *Erasmus Darwin* (Charles Scribner's Sons, New York, 1963), p. 114.

13 Linnaeus, *Praeludia*, cited in Larson, 'Linnaeus and the Natural Method,' *Isis*, 58 (1967), pp. 304–20, esp. p. 306.

14 *Oxford English Dictionary*, s.v. 'gamete'.

15 Blunt, *The Compleat Naturalist*, pp. 176–7.

16 Ibid.

17 Bernard de Jussieu to Linnaeus, 30 Jan. 1749, in *The Correspondence of Linnaeus*, ed. Smith, p. 221.

18 Blunt, *The Compleat Naturalist*, pp. 165–6.

19 On style in scientific language, Wolf Lepenies, *Das Ende der Naturgeschichte: Wandel kultureller Selbstverständlichkeiten in den Wissenschaften des 18. und 19. Jahrhunderts* (Suhrkamp, Frankfurt, 1978); Ludmilla Jordanova (ed.), *Languages of Nature* (Rutgers University Press, New Brunswick, 1986), and Londa Schiebinger, *The Mind Has No Sex? Women in the Origins of Modern Science* (Harvard University Press, Cambridge, Mass., 1989), chapter 5.

20 Cited by Ronald King in Robert Thornton, *The Temple of Flora* (1799; New York Graphic Society, Boston, 1981), p. 19.

21 Linnaeus loved especially the great poets of Rome – Virgil and Ovid. In addition to classical mythology, Linnaeus also used images from traditional wedding poetry and the Book of Psalms of the Bible. See Sten Lindroth, 'The two faces of Linnaeus,' in Tore Frängsmyr (ed.), *Linnaeus: The Man and His Work* (University of California Press, Berkeley and Los Angeles, 1983), p. 10. See also John L. Heller, 'Classical Mythology in the *Systema Naturae* of Linnaeus,' *Transactions and Proceedings of the American Philological Association*, 76 (1945), pp. 333–47.

Linnaeus's sexual images came most immediately from France – especially from the botanist Sébastien Vaillant, whose essay on plant sexuality influenced him as a young man. But that is the subject of another essay.

22 Anna Seward, *Memoirs of the Life of Dr. Darwin* (Philadelphia, 1804), p. 95. Darwin also found the Rosicrucian doctrine of gnomes, sylphs, nymphs, and salamanders proper machinery for a botanic poem. He saw hieroglyphics as embodying in allegory many truths of nature known to the Egyptians which were later taken over into the mythology of Greece and Rome. Erasmus Darwin, *The Botanic Garden* (1791; Garland Publishing, New York, 1978), Apology, pp. vii–viii.

23 It has been argued that Linnaeus was essentially a scholastic who ordered the world of plants into classes, orders, genera, and species along the same lines as the learned doctors of Middle Ages. Lindroth follows Julius Sachs in this interpretation: see Lindroth, 'The Two Faces of Linnaeus,' pp. 33–7.

24 Allen, *The Naturalist in Britain*, chapters 1 and 2.

25 John Miller, *An Illustration of the Sexual System of Carolus von Linnaeus* (London, 1777); James E. Smith, tr., *A Dissertation on the Sexes of Plants* (London, 1786); and Robert Thornton, *A New Illustration of the Sexual System of Carolus von Linnaeus* (London 1799–1807).

26 Charles Alston, 'A dissertation on the sexes of plants,' *Essays and Observations Physical and Literary Read before the Philosophical Society in Edinburgh* (1754; Edinburgh, 1771), vol. 1, pp. 315–18.

27 William Smellie, 'Botany,' *The Encyclopaedia Britannica* (Edinburgh, 1771), vol. 1, pp. 627–53; and *The Philosophy of Natural History* (2 vols, Edinburgh, 1790), vol. 1, pp. 245–63. In 1790 John Rotheram published

The Sexes of Plants Vindicated as a reply to Smellie's scientific objections, but he did not mention the matter of morals.

28 Smellie, 'Botany,' p. 653.

29 William Withering, *A Botanical Arrangement of British Plants* (2 vols, Birmingham, 1787), vol. 1, p. xv. Ludwig of Leipzig also avoided sexual terms in his explication of Linnaeus's classes and orders by substituting the straightforward botanical terms 'anther' and 'style' for Linnaeus's 'andria' and 'gynia.' Linnaeus's first class was thus rendered 'monatherae' and his first order 'monostylae,' etc. Richard Pulteney, *A General View of the Writings of Linnaeus*, 2nd edn (London, 1805), p. 243.

30 Johann Siegesbeck, *Verioris brevis sciagraphia* (St Petersburg, 1737), p. 49. Linnaeus, who never took kindly to criticism, named an unpleasant small-flowered weed *Siegesbeckia*; see King, in Thornton, *The Temple of Flora*, p. 9.

31 Amman to Linnaeus, 15 Nov. 1737, in *A Selection of the Correspondence of Linnaeus*, ed. Smith, vol. 2, p. 193.

32 Darwin, *The Loves of the Plants*, canto I.

33 Erasmus Darwin, *Zoonomia* (London, 1794), vol. 1, p. 147.

34 Seward, *Memoirs of the Life of Dr. Darwin*. For a different view see, King-Hele, *Erasmus Darwin*, pp. 116–17.

35 James Logan, 'The poetry and aesthetics of Erasmus Darwin,' *Princeton Studies in English*, 15 (1936), p. 114.

36 *The Monthly Review*, 9 (June 1793), p. 183. The French translator of Darwin's poem gives a genealogy of botanic poetry stretching from Ovid to Paul Contant's 1609 *Jardin et cabinet poétique*: see J.-P.-F. Deleuze, *Les amours des plantes* (Paris, 1800), preface.

37 Letter to James Watt, 20 Nov. 1789, reprinted in *The Letters of Erasmus Darwin*, ed. Desmond King-Hele (Cambridge University Press, Cambridge, 1981), p. 196. Darwin sold his *Loves of the Plants* for £800 and his *Botanic Garden* for £900: see *Letters*, pp. 206, 225.

38 Letter to Joseph Johnson, 23 May 1784, ibid., pp. 139–40.

39 Seward, *Memoirs of the Life of Dr. Darwin*, p. 95.

40 Darwin, *The Botanic Garden*, vol. 1, p. v.

41 Letter to James Watt, 20 Nov. 1789, *Letters*, p. 196.

42 King-Hele, *Erasmus Darwin*, pp. 116–17.

43 Seward, *Memoirs of the Life of Dr. Darwin*, pp. 124–5, 157. In the whole of *The Botanic Garden*, Seward found only one objectionable passage, that of a tale retold from Homer, in which azoic gas is personified as Mars and made lover to the virgin air (Venus) with fire (Vulcan) the jealous rival. Here Darwin, Seward judged, has overstepped – like Homer before him – the bounds of propriety with his libertine and unjust deities.

44 Maureen McNeil, *Under the Banner of Science: Erasmus Darwin and His Age*, (Manchester University Press, Manchester, 1987), pp. 64–6.

45 See his celebration of the French Revolution in his *Economy of Vegetables*, IV, pp. 341–4.

46 Norton Garfinkle, 'Science and Religion in England, 1790–1800: the critical responses to the work of Erasmus Darwin,' *Journal of the History of Ideas*, 16 (1955), p. 378.

47 Ibid., p. 380; see also McNeil, *Under the Banner of Science*, pp. 133–5.

48 Roy Porter, 'Mixed feelings: the Enlightenment and sexuality in eighteenth-century Britain,' in Paul-Gabriel Boucé (ed.), *Sexuality in Eighteenth-Century Britain* (University of Manchester Press, Manchester, 1982), pp. 1–27.

49 Ibid., pp. 8–15, 19–20.

50 McNeil, *Under the Banner of Science*, pp. 80–4. See also, William Withering's account in *The Miscellaneous Tracts* (2 vols, London, 1822), vol. 2, pp. 115–23.

51 *The Anti-Jacobin: or, Weekly Examiner*, 23 (April 1798), pp. 180–2; 24 (April 1798), pp. 188–9; and 26 (May 1789), pp. 204–6. Darwin apparently pretended that he had never heard of this satire. Geometry may have been chosen as the satiric vehicle in this case because geometric lines were at this time distinguished into genders, classes, or orders, according to the number of dimensions of an equation expressing the relation between the *ordinates* and the *abscissae*: *Dictionarium Britannicum*, ed. G. Gordon, P. Miller, and N. Bailley (London, 1730), s.v. 'gender.' The same dictionary defines 'man' as a creature endowed with reason, and though an etymology is given for woman (pamb or womb plus man), 'woman' is not defined.

52 Reverend Richard Polwhele, *The Unsex'd Females; a Poem* (1798; New York, 1800), pp. 33–4.

53 Ibid., pp. 10–20.

54 Polwhele took Emma Crewe to task for overstepping the modesty of nature in her portrayal of Cupid in the frontispiece to Darwin's *Loves of the Plants*: ibid., p. 27.

55 Erasmus Darwin, *A Plan for the Conduct of Female Education in Boarding Schools* (London, 1797). Darwin presents traditional images of women; absent from his botanic reverie are personifications of radical or intellectual women like Mary Wollstonecraft. See Janet Browne, 'Botany for Gentlemen: Erasmus Darwin and *The Loves of Plants*,' *Isis*, 80 (1989), pp. 593–621.

56 See Desmond King-Hele, *Doctor of Revolution: The Life and Genius of Erasmus Darwin* (Faber and Faber, London, 1977), pp. 234–7. King-Hele has incorrectly emphasized the radical nature of Darwin's tract.

57 Robert Thornton, *The Politician's Creed* (London, 1795–9), title-page and introduction.

58 Carl Linnaeus, *Sponsalia plantarum* (1746) in *Amoenitates academicae* (Erlangen, 1785–9), p. 329.

59 *Dictionarium Britannicum*, s.v. 'pistillum.' J.-P. de Tournefourt standardized

the modern usage of the term pistil (*pistile*) to designate the ovary, style, and stigma of the flower in 1694; the term was introduced into botanical usage in England in the 1750s. The use of the term 'stamen' is much older; Pliny referred to the stamen of the lily. The technical use of the word in botany, however, dates to about 1625.

60 See Schiebinger, *The Mind has no Sex?* chapters 7 and 8.

5

Dichotomy and Denial:
Mesmerism, Medicine and
Harriet Martineau

Roger Cooter

Until very recently it was scarcely permissible for historians to focus on individual accounts of women's experiences of illness and recovery. Cultural historians looked askance at narratives of any sort; social historians of science and medicine set themselves against antiquarian biographical traditions; and feminist historians – appreciating that the 'rediscovery of the worlds that women have inhabited – can lead to a ghettoization of women's history'[1] – had all the more reason to turn away from the somatic or psychic states of particular women.

But nothing changes like history itself. Although there has been no intellectual retreat, agendas have shifted. Within the 'new historicism', for feminists especially, the study of 'narrative logic' and the relationship between texts and contexts has become all important; at least two recent works have come to focus on narrative accounts of health, sickness and suffering in history in order to piece together broader cultures of illness-experience; while studies such as that by Julia Epstein on Fanny Burney's account of her mastectomy of 1811 serve in part to illustrate deployments of professional authority against female autonomy as symbolized by the sacrosanct female body.[2] Following Foucault and others, of course, the subject of the female body itself has become a central historical preoccupation.[3]

This chapter, although it takes up a medical narrative, is not concerned with the study of narrative as such. Nor is it about cultures of female healing, though with these it partly intersects. Nor, finally, is it intended primarily as a contribution to the study of its central character, Harriet

Martineau (1802–76). Rather, this chapter draws on Martineau's unique experience of illness and cure in order to contextualize and comment on the construction and reconstruction of gender in medicine between the 1840s and 1870s. For this, the intimate history of Martineau's body is unusually revealing: her medical notoriety was based on what was defined as a gynaecological problem, her experience of cure was through the use of mesmerism, and her case prompted extensive commentary both in the 1840s and in the 1870s.

The potential of this historical episode rests on more than just the body in question, however; it rests, too, on the stature of the woman herself. Harriet Martineau became famous in the 1830s through her popularizations of bourgeois political economy, and this reputation was sustained through a life of unceasing journalistic interventions into the major social and intellectual events of her time. Among those interventions, besides an advocacy of mesmerism, was an espousal of liberal views on women's place in society. Yet the latter is only a part of the reason why (since the early 1970s) Martineau has come to occupy a place of honour between Mary Wollstonecraft and Simone de Beauvoir in feminist anthologies.[4] As significant is the fact that to her contemporaries (male and female alike) she was often regarded as 'one of the finest examples of a masculine intellect in a female form which . . . [has] distinguished the present age.'[5] As such, Martineau's life-experience mocks simplistic reductions, especially to the ideology of 'separate spheres'. Instead, her spinster life reveals fundamental conflicts between female interests and ambitions and dominant Victorian prescriptions for women's role and function. Thus, as one feminist biographer has recently insisted, it is necessary to study Martineau's life dialectically, in terms of 'resistance to and complicity with the hegemony of male dominated middle-class culture.'[6]

This chapter adopts this revisionist perspective in its exploration of how Martineau's experience with mesmerism can be seen historically both to conspire with and resist one-dimensional accounts of medicine's role in the construction of gender relations. Although, as we shall see, to some extent the medical profession was ultimately to 'one-dimensionalize' Martineau's reality, its difficulties in doing so serve to expose realities that were otherwise. Hence, for us, her experience can serve to challenge those historical accounts of the relations between medicine and gender that, in their relentless search for a uniform history of women's oppression by men, similarly overlook complexity, variation and contradiction.[7]

MEDICINE V. MARTINEAU

Harriet Martineau has left a multitude of primary material which illuminates the contradictions of her experience of medicine. Besides her own published accounts and private letters describing the state of her body before and after her cure by mesmerism, there exists a detailed clinical account, published in 1845 by her physician and brother-in-law, Thomas Greenhow, as well as a published account of the post-mortem that was conducted in June 1876 by her last physician, W. Moore King. There are also reports by the distinguished physicians she consulted, among them the heart specialists Peter Mere Latham and Sir Thomas Watson. Additionally, there are extensive reviews of her case in the medical journals. Such texts provide contradictory and compelling accounts of Martineau's body and medicine's construction of it.

According to her own account, her first intimation of any 'female problems' occurred when she was thirty-seven. Though always sickly, and partially deaf from youth, it was while she was in Venice in 1839 that she became sensible of a 'great failure of nerves and spirits, and of strength', with menstruation 'occurring every two or three weeks' accompanied by an irritating discharge of brown and yellowish fluid. Shortly thereafter she found 'a solid substance' protruding from her vagina, 'in form resembling the end of a bullock's tongue, with a decided edge or point.' Before long she could neither insert a syringe nor sustain a sponge pessary, which convinced Greenhow – to whom she communicated all this – that the problem was either a prolapsed uterus or a 'polypus tumour'.[8] Additionally, Martineau was troubled by constipation and nervous disorder.

When she returned to England and placed herself under Greenhow's care in Newcastle in July 1839, he was able to confirm the displacement of the uterus, but not the tumour, which he suspected had now slid up into the pelvic cavity. Anticipating that it might reappear, he in the meantime ordered warm baths, ergot of rye and opiates, and had leeches applied to relieve the tenderness in the left groin. But there was no change in Martineau's condition, and when the famous accoucheur, Sir Charles Mansfield Clarke, examined her in September 1841, it was only to confirm Greenhow's diagnosis of a non-specific 'enlargement of the Body of the Uterus', and to add to Greenhow's regimen external applications of iodine ointment.[9]

And so, in considerable pain – apparently much worsened by the iodine ointment – Martineau resigned herself to an invalid's life on the sick-couch. Although she continued her literary labours,[10] to the extent that she was now prostrate and ruled over by her reproductive organs, she came to epitomize the patriarchal rendering of the Victorian woman. With her mind trapped in an idealized female body, as it were, she wrote *Life in the Sick Room* (1844) in which she adorned the female stereotype by extolling the merits of illness in strengthening Christian fortitude.[11] The reasons why Martineau entered into this caricature of the feminine sex role doubtless owe as much to her personal psychology as to the contemporary sociological situation of women. It is known that Martineau was deeply scarred by the gender role-limitations set by her mother, and it has recently been suggested that by taking to her couch and allowing her femininity to be defined for her she was at the same time liberating herself from her past, or coming to terms with her own femininity and moving 'towards a sense of identification and community with other women.'[12]

For the moment, however, we need only note that Martineau remained in her prostrate condition – 'preserving a decent quietness in the midst of our troubles'[13] – until June 1844 when, at Greenhow's suggestion she eagerly volunteered to be mesmerized by the itinerant phreno-mesmerist Spencer Timothy Hall, who was then in Newcastle. Hall was in fact unable to mesmerize Martineau, and it was her maid, Jane Arrowsmith, mimicking Hall on an occasion during his absence, who eventually succeeded in putting Martineau into a trance. Almost immediately her condition improved. By September 1844, after repeated mesmeric treatment by Arrowsmith and other female mesmeric healers, Martineau found herself returned to a life she had long ceased to expect. At the age of forty-two her menstrual cycle became regular; she was able to leave off the use of opiates, and her chronic constipation, gastric problems and nervous disability declined. Through an occult means, whose power she found worked unexpectedly and 'was wholly unlike anything I have ever conceived of', she was able to move about out of doors for the first time in years. In-valid no more – indeed, a different person in a different body – she rushed into print to proclaim the 'vast and mysterious' powers of mesmerism.[14]

Martineau and Mesmerism

Mesmerism was at this time at the height of its popular success, much aided and abetted by stage performances of phreno-mesmerism, such as those by S. T. Hall.[15] But because of its 'rapid progress and its singularly enthusiastic reception among the excitable and untutored masses',[16] the subject was anathema to most medical men, especially to those, such as Thomas Wakley of *The Lancet*, who were earnestly seeking to enhance their own and the profession's status. Only a few years before (in 1838), the leading British medical advocate of mesmerism, John Elliotson, had been forced to resign his professorship at University College, London, after Wakley had exposed as 'fraudulent' the clairvoyant powers of Elliotson's prize performers, Jane and Elizabeth Okey.[17] To reformers like Wakley, homoeopathy, hydropathy and botanic medicine were bad enough, but mesmerism was worse. Apart from the economic threat it posed in the open medical market, its alleged sexual dangers, and its availability to the 'untutored masses', it challenged the very foundations of the 'rational' scientific knowledge with which they sought to be aligned. Mesmerism, it had long been maintained, was 'without analogy'; its 'essential qualities' were seen as 'too rare and subtle ever to be made the subject of a natural experiment.'[18] But more than this, mesmerism beat orthodox medicine at its own game: first, by purporting to abolish pain and suffering at literally a few strokes, and, second, by enabling certain persons to become clairvoyant and thereby divine the internal ailments of others. The latter had been the claim of the Okey sisters, which Wakley had managed to 'expose'; but it was to be another few years before the former claim would be deflated.[19]

The timing of Martineau's mesmeric cure could hardly have been worse. For it was precisely at this point that the medical reformers were striving to establish a consensus on the dangers of unlicenced quackery in order to legitimate their interests against those of the old medical establishment associated with the Royal Colleges. Indeed, on one occasion, the *Lancet* reported on the reformers' protest against the government's 'total inadequacy to protect the rights of the qualified practitioner', on the same page that it reported on a Fellow of the Royal College of Surgeons found guilty of giving 'fresh impulse to mesmerism . . . since the publication of Miss Martineau's case.'[20] Instead of enabling the medical reformers to exploit 'the dangers of mesmerism' as a resource in professionalization, the Martineau case worked powerfully in the opposite direction, exposing what Martineau and many others came to believe was 'the apathy, cowardice or mercenary spirit of the Medical Profes[n].'[21] Martineau, alas, was quite unlike the young Okey sisters, who were ostensibly silly, impressionable and working class. Her reputation went before her, not only as a mature, supremely sensible, 'man-like' rationalist, but also as a bourgeois radical whose ideological outlook was about as near as could be hoped to the medical reformers themselves. As the *Lancet* lamented in an editorial: 'If Miss MARTINEAU had been considered a quack in politics and literature, her present performance might have been regarded as unworthy of attention.'[22] As it was, news of her mesmeric cure was like a frontal assault from a fifth columnist; the *Letters on Mesmerism* (1845) betrayed only too well how an intelligent woman's circumspection about mesmerism had been overcome by proofs of its efficacy. The worst of it, of course, was that the *Letters* were immediately taken up both among 'the popular' *and* 'the polite'. As John Elliotson was delighted to report,

The subject which the critic, a few months since, would not condescend to notice, has been elevated to a commanding position. It is the topic with which the daily papers and the weekly periodicals are filled; in fact, all classes are moved by one common consent, and mesmerism, from the palace to the smallest town in the United Kingdom, is the scientific question absorbing public attention. . . . The immediate cause of all this activity is the publication of the case of Miss Martineau, who, after five years' incessant suffering and confinement to her couch, is now well.[23]

Elliotson had additional reason to be excited about Martineau's cure, for the great dispute at this time was whether mesmerism could cure *organic* disease, as opposed to merely psychological conditions.[24] Since Martineau's case indicated that mesmerism apparently could perform such cures, the basis for the dismissal of mesmerism through association with 'weak-minded' suggestive females was further weakened.

It might have been expected, though, that the orthodox medical press would have sought to counter this claim. After all, hysteria, or hysteria's little cousin 'nerves', was a conventional way of belittling non-passive female behaviour, and the knowledge in this case of the specifically uterine nature of the problem rendered this type of explanation all the more likely. Such, indeed, was the line of argument pursued by the author of a thirty-three page review of Martineau's *Letters on Mesmerism* in the *Edinburgh Medical and Surgical Journal* in 1845. After also drawing attention to the past links between mesmerism and political extremism, the author went on to remark:

> In the present narrative . . . there is nothing peculiar. A lady labours under disease in the uterus, a circumstance of daily occurrence. She gets out of health, – also a common occurrence. She becomes nervous, fanciful, feeble in body and imbecile in mind, – also common occurrences. . . . Partly under the disease and partly under the treatment, her mind becomes still more enfeebled. She is nervous, hysterical, cataleptic, somnambulous; and in this condition is visited by a mesmeriser, who kindles up into active conflagration the smouldering fancies of her distempered brain. In all this there is nothing new; nothing extraordinary, and nothing wonderful. . . .
> The case of Miss Martineau is not in any respect more wonderful than the cases given by Joseph Frank, the cases of those nice young women, the Okeys, or the cases of any other half-dozen, or half-a score of hysterical females.[25]

That the reviewer had to work so hard to establish that there was nothing extraordinary in the case only suggests that to most contemporaries, the opposite was probably more apparent. Although most of Martineau's modern biographers are inclined to the view that some degree of hysteria *was* involved in the case, and feminists in particular have accepted this view as confirming Carroll Smith-Rosenberg's well-known characterization of hysteria as a rebellion against female gender requirements,[26] most of the contemporary medical press chose not to dwell on it. Typically, the *Medico-Chirurgical Journal* hoped only that the case would 'be the wonder of the day – perhaps of nine days – and then sink into oblivion with the exploits of Miss Okey.'[27] The *Lancet*, of course, knew better; indeed, this was the first occasion for its mentioning mesmerism since its total ban on the subject after the affair with Elliotson in 1838. But what was then an apoplectic response, became now calculated prudence. The journal serialized a moderate 'History of Mesmerism' between February and May 1845, which tried to show both that there was nothing new in mesmerism, and that the existing evidence for it was insufficient and unscientific, if not bogus. Within this account, the Martineau case was referred to only in passing, and hysteria likewise.[28]

Although the main medical document relating to the Martineau case, Thomas Greenhow's *Medical Report of the Case of Miss H —— M ——*

(1845), *did* draw attention to the role of 'the morbid influence over the nervous system', this was regarded as merely 'additional' or secondary. Written ostensibly because of 'the general interest which [the Martineau case] has excited', but in reality to save Greenhow's own reputation and to clear his name among his professional brethren for having introduced Martineau to mesmerism, the *Report* sought to stress that for some three months *before* Martineau encountered mesmerism, 'slight changes' were detectable in the condition of her uterus. The alleged changes were claimed to be the start of a wholly explicable return of Martineau's uterus to its normal position.[29] This account – the intimate nature of which within a shilling pamphlet led Martineau to sever all relations with Greenhow – was disputed by S. T. Hall among others, who claimed that the explanation was *post hoc*.[30] The important point here, however, is that Greenhow, far from seeking to attribute much to female hysteria, endeavoured by reference to pathology to deny altogether that mesmerism had anything to do with the cure, if cure there were.

Like Martineau, Greenhow was a Unitarian who was closely associated with radical bourgeois reformers within medicine and without. In the 1830s he had been the chairman of the wholly middle-class, reformist, Newcastle Phrenological Society.[31] Like Wakley, who was also an advocate of phrenology, Greenhow was in no position to tar Martineau's mesmerism with the brush of political extremism – least of all with antifeminist strokes. He could hardly have been unaware that in *Society in America* (1837) Martineau had deplored the fact that 'indulgence is given [women] . . . as a substitute for justice'; that women were degraded into the position where marriage was 'the only object left open to [them]'; and that, as concerns women's rights, even the best politicians and thinkers were despots – 'Jefferson in America, and James Mill at home, subside, for the occasion, to the level of the author of the Emperor of Russia's Catechism for the young Poles.'[32] Like virtually every exponent of women's rights since at least as far back as Mary Astell in 1696, Martineau also drew the parallel between the position of women in society and that of plantation slaves.[33] All of this made it rhetorically difficult for advanced Liberals (and anti-slavery advocates) like Greenhow and Wakley to attack Martineau's defence of women's rights. Nor was it politic for them to do so in the light of the fact that it was Tory journals, such as the *Quarterly Review* in 1844, that were commenting on 'the subject of "Woman" ' as being 'so terrifying'.[34] Although Greenhow and Wakley might have wanted to agree with the author of the article in the

Edinburgh Review in 1841 who, surveying the spate of recent works on women's education, rights and conditions, concluded that the idea of raising women to the level of men was 'plausible but unsound',[35] in the absence of unequivocal scientific and medical proof it was unwise for them to take it up. Phrenological evidence on the matter was at best ambiguous, and socialists and feminists were known to be using such evidence to legitimate social and sexual equality.[36]

In so far then as we might see two processes operating in the medical discussion of Martineau and mesmerism in the 1840s – the one, demarcating medicine from quackery, or professional from lay views, and the other, relying on hysteria to substantiate distinctions between males and females – the latter appears weaker largely for political professional reasons. Overall, the Martineau case served medical interests poorly, while reinforcing the belief of advocates of mesmerism that the phenomena was not 'confined . . . to one sex, nor to any particular age or class nor to any country.'[37] To a degree, Martineau's 'mannishness' only further proved the case.

POST-MORTEM

But if the medical reaction to Martineau's mesmeric cure in 1840s reflects mainly on the profession's insecurity in relation to unorthodox medicine, to what can we attribute the fact that in 1876, upon Martineau's death, the subject of her mesmeric cure once again preoccupied the medical press? By this date, not only had the Medical Registration Act been in force for some two decades, but lay mesmeric healers had long ceased to be a serious threat to the profession. If the profession had had cause to worry that Martineau's example might lead other women with uterine problems to seek help from mesmerists,[38] this worry had subsided as the profession itself appropriated and scientized mesmerism in the form of hypnosis.[39]

On the face of it, 'sweet revenge' might seem explanation enough for the profession's renewed interest in the case. Certainly for Thomas Greenhow, who was then in his mid-eighties, it was worth coming out of retirement in order to vindicate his claims of 1845.[40] For the post-mortem apparently gave ground to his and Clarke's speculation on the possibility of a tumour as the cause of the prolapse and its retroversion. Martineau's illness, it was now claimed, had been the result of the growth of a benign left-sided ovarian cyst. Initially, so the claim went, the cyst had caused her uterus to prolapse, but upon the subsequent growth of the cyst the uterus had been forced back up into the pelvis – the latter process apparently coinciding with the initiation of the mesmeric treatment. It was speculated that upon the cyst's further growth, the stomach had been forced up into the thoracic cavity, arching the diaphragm and seriously interfering with the action of the lungs and heart.[41] Thus was explained Martineau's conviction in the mid–1850s that she was suffering from a heart condition. She was certain that death was imminent from this cause and wrote her autobiography in great haste. When death finally came, two decades later, the cyst had swollen to a capacity enough to hold five quarts of liquid, leaving Martineau in her last years looking distinctly like an elderly pregnant woman. Once again her body itself seemed to flaunt normative boundaries, as if in rapport with her mind.

Despite the size of the cyst, however, Martineau's morbid pathology was hardly so extraordinary as to merit the minute and extensive coverage that it received in the medical press, let alone the interventions of the country's then leading ovariotomist, Thomas Spencer Wells. Dangling the distended ovary before the Clincial Society of London in April 1877, Wells depicted how 'as the organ grew larger, it [rose] as if from the kitchen to the parlour, intimating that relief would be then obtained, but not cure.'[42] Such a metaphor, so resonant of hierarchies of class, gender and domestic production and consumption, is too rich to elaborate here, but its dominant notion of crossing established boundaries is worth noting. As far as Wells was concerned, however, such description was only by way of proving that his professional brethren, Greenhow and Clarke, had been wholly correct in their diagnoses, and that so, too, had been the specialists, Latham and Watson, in dismissing Martineau's fears of a heart condition.

But what was the need? As a commentator in the *British Medical Journal* put it in November 1877:

> At this time of day, Miss Martineau, one would think, might be pardoned her offences against medicine, in concerning herself with mesmerism; and especially since time . . . had in this case left medicine master of the position, *victor juvansque*. . . . [N]eed we reproach Miss Martineau if, on such a sudden resurrection of life and comfort, she should have confused the *post hoc* with the *propter hoc*?[43]

A part of the explanation lies in the fact that shortly after the medical profession opened up Martineau, the public was cutting open the pages of

PLATE 5 Harriet Martineau, (*c*.1867),
(Reproduced by kind permission of the National Portrait Gallery, London)

her posthumously-published *Autobiography* (three editions of which appeared in 1877 alone). Written in 1855, the *Autobiography* reflected Martineau's then far-from-friendly attitude to the medical profession. Whether or not she was really expecting 'that the whole system of Medicine . . . [was to go] to the dogs . . . and that henceforth, instead of Physicians, we are going to have Magnetisers!' (as Jane Carlyle reported sarcastically); there is no doubt about her keenness 'to show what treatment medical men inflict on women of my rank who have recourse to mesmerism.'[44] As the *Letters on Mesmerism* had previously made clear, it was in reaction to such treatment that she had gathered around her a coterie of female mesmeric healers, who undertook the care and cure of each other, as well as conducting healing activities within the wider community. Martineau's own mesmerist, Mrs Montague Wynyard, was a Londoner, who came to live with her in Tynemouth; and in her serving girl, Jane Arrowsmith, Martineau claimed to have found a remarkable clairvoyant. Consciously or unconsciously, these women inverted the usual naked reproduction of gender relations in the induction of mesmeric trances – the surrender, passivity and submission which also epitomized the usual female-patient/doctor relationship. Significantly, the only man involved in these healing activities was Henry Atkinson, Martineau's doted-upon 'mesmeric adviser', who was almost certainly homosexual, thus violating gender norms and behaviour himself.[45] Yet it should also be noted that in the use of the serving girl as the person who had to submit most frequently to the dominating psychic power of the mesmeric operators, there was the reproduction of existing class relations. Moreover, traditional class and gender roles were reproduced by Martineau herself once she acquired the power of mesmeric healing. The rustics around her new home in the Lake District became the objects of her charitable healing and, on one occasion, so too did her ailing cow 'Betsy'.[46]

While the class dimensions to Martineau's use of mesmerism might be overlooked by the medical profession, the challenge to their authority symbolized by her endorsement of mesmerism could not. The discussion on the post-mortem in the *British Medical Journal*, which extended to December 1877, was in fact initiated in July 1876 by rumours of a posthumous publication on mesmerism by Martineau, and by a brief report on her obituary in the *Daily News*. The latter was seen as containing 'some very painful matter for the professional [medical] reader', and as instancing a case which 'is not one of the triumphs of the profession'.[47] Martineau had written the obituary herself and, as in her *Autobiography*, she had stressed her complete cure by mesmerism in 1844 in the face of eminent medical opinion declaring her case impossible of recovery. She also stressed that her death (now twenty-one years after the medical profession had declared it imminent), was wholly a result of a 'deterioration and enlargement of the heart'.[48] Martineau in fact knew this not to be true; after the consultations with Latham and Watson in 1855, she secretly acknowledged 'the tumor' as the source of her ill health.[49] But by denying this, and purporting to die from a different illness, she was able to continue to proclaim the curative power of mesmerism. At the same time, by professing a 'heart condition' she escaped the trap of female illness, and was thus empowered in the male world with a 'respectable' illness capable of being openly discussed.

In the light of her posthumous claims, it is easy to understand why the medical profession should have been anxious to set the record straight as to the 'real facts' of Martineau's illness, supposed mesmeric 'cure', and eventual death. But the involvement of gynaecologists points to the deeper implications of Martineau's case. Perhaps the fact that nothing was made explicit at the time about this section of the medical profession's need to set the record straight can itself be taken as significant. In what follows I want to explore this possibility by briefly elaborating some of the circumstances surrounding gynaecology in the years between Martineau's mesmeric cure and her post-mortem, and by relating these to the broader politics of gender involved in Martineau's life story.

MARTINEAU AND GYNAECOLOGY

At the same time that Martineau was experiencing her mesmeric cure, 'gynaecology' was becoming one of the most active theatres of medical professionalism.[50] Whereas in 1814 Charles Mansfield Clarke could observe in the preface to his influential *Observations on Those Diseases of Females Which are Attended by Discharges* that 'the diseases of the sexual organs in females . . . are perhaps less generally known and understood by practitioners, than any other complaints to which the human body is subject', by 1846, as the *Lancet* then put it, to 'write a book (no matter how flimsily) on . . . diseases of women attended by discharges' had become one of the commonest modes of cutting a reputation in medicine.[51] Obviously self-interest was crucial to this involvement; the medical profession was then overcrowded, and treating the illnesses of middle- and upper-class women was lucrative.[52] However, precisely because a good reputation was needed to successfully enter into this increasingly competitive market, the 'reputation-earning' gynaecological texts seldom strayed from being narrow, if not always 'flimsy', clinical accounts of female pathology. Occasionally the stated motive was that of preventing women from frequenting charlatans and medical empirics,[53] but most often these works referred in their prefaces only to the need for 'more minute investigation[s] into the phenomena of the healthy performances of the functions peculiar to women' in order to throw light on the causes of disease and promote their prevention and cure.[54]

Contrary to what some feminist historians sought to argue in the 1970s,[55] little of this literature was driven overtly by the need to legitimate male superiority and politically oppress women, let alone specifically to reassert traditional patriarchal values in the manner of earlier texts on the 'duties of women'.[56] This is not to say that the proliferation of these texts cannot, or should not, be read in terms of the politics of gender. On the contrary, they must be read in this way since fundamental to them was the abstraction or reification of women as subjects for male intellectual inquiry. As such, they can be located within a tradition extending at least as far back as Dr William Alexander's *The History of Woman from the Earliest Antiquity to the Present Time* (1779) (which had been explicitly concerned to prove the natural equality of the sexes),[57] and at least as far forward as the late nineteenth-century publication by the gynaecologist Herman Ploss, *Woman: An Historical, Gynaecological and Anthropological Compendium*, in which 'woman' was conceptualized wholly as if an exotic species of plant.[58] The clinical literature might be said to have differed from both these volumes only in so far as its abstraction of women was usually unaccompanied by explicit social commentary. Indeed, one of the outstanding features of this literature between the late eighteenth and early nineteenth centuries, as the historian of gynaecology James Ricci noticed in 1945, was its elimination of overt social categories.[59] Whereas, 'in a vague and ill-defined manner' female maladies in the eighteenth century had been grouped according to those common to women in general, those common to virgins, those common to married women, and those common to widows, by the early nineteenth century the uterus alone had come to predominate. Thus within this clinical literature fundamental social categories were collapsed into a single pathological one.

But there was more to it than this, as several recent scholars have made clear.[60] Hand in hand with the pathological reduction of woman to her uterus went the view that female biology was not parallel to that of males. Before the early nineteenth century, it had been common to regard the female reproductive organs as analogous to those in men; biologically women were men turned outside in. Thereafter, however, the uterus came to be construed as something without analogy in men; women and men became biologically incommensurable. Thus the relationship between the sexes was no longer to be one of equality or inequality grounded in politics, but, rather, one of difference grounded in biology. As Auguste Comte was to state in the fourth volume of his *Cours de philosophie positive* (1839): 'The sound philosophy of biology, especially the important theories of [the phrenologist] Gall, begin to offer a scientific resolution to the much acclaimed equality of the sexes. The study of anatomy and physiology demonstrates that radical differences, at once physical and moral . . . profoundly separate the one [sex] from the other.'[61] Increasingly, within the gynaecological literature, women were to be seen as inferior because they had faulty parts which men simply did not have. 'Faultiness', moreover, became the quintessence of the female condition. For the uterus was now to be seen as the organ governing the entire female organism whether a woman was pregnant or not.[62] To lesser or greater degrees all women were hysterics, perpetually ruled over by their reproductive organs. They were diseased by virtue simply of being women.

Gynaecology itself was not, of course, responsible for this change. Science and medicine are not autonomous; they do not make power-relations so much as embody, mediate and capitalize upon them. The development of gynaecology is merely the transformation of men's power over women into science. But this is not to underestimate the consequences; after all, what had formerly been resisted by feminists as merely the 'tyranny of custom', or 'as much the work of art as of nature',[63] now emerged out of gynaecology as an irresistible fact of nature. Women were doubly disadvantaged by this, for gynaecology's scientization of gender relations was also a part of the general trend in medicine in these years to denigrate lay views of the body and illness experience. Gynaecology came both to epitomize male power in medicine *and* to epitomize the assertion of professional over lay views. Put otherwise, inherent to the pathologization of women was the denial of women's own account of their illness experience. Sir Charles Mansfield Clarke, for example, maintained that for a physician to treat female diseases on the basis of the patient's own perception of her illness was comparable to expecting 'to remove ascarides [round worms] from the rectum by making application to the nostrils'.[64] This was Clarke's view in 1814, in the so-called 'Dark Ages of uterine pathology', before Recamier demonstrated the possibility of ocular examination of the womb.[65] By 1842, when Martineau was expressing her dissatisfaction at Clarke's minute examination of her vagina (heartily wishing 'he had never come northeast'),[66] Clarke and his colleagues were on the eve of fulfilling Recamier's ambition: in 1843 James Simpson and P. C. Huguier independently developed uterine sounds and curettes with which to extend the medical gaze beyond the reach of the speculum, up the cervical canal and into the uterus itself.[67] Conditions that previously had only been detectable through 'discharges', physiognomical signs and other outward appearances – appearances as open to the readings of patients as physicians – were now thoroughly mystified, and the power of detection vested entirely in the hands of gynaecological experts.

Thus, at precisely the time that Martineau's cure by mesmerism was becoming public, gynaecology was becoming *the* place to exemplify not only the superiority of the (would-be scientific-rational) medical gaze over the lay view, but also the biological superiority of males over females. The expression of the one through the other, or the clarification of the latter through the former, rendered gynaecology doubly powerful in disadvantaging women. However, in the light of the previously mentioned professional political circumstances surrounding the Martineau case in the 1840s, it is perhaps not surprising that its discussion should then have been primarily in terms of the boundaries between orthodox and heterodox medicine, or professional and lay. Thereafter, it might be said that the polarities of sex and gender were allowed to overlay those of orthodox/fringe, or professional/lay. As the work of Cynthia Russett most recently reminds us, during the last third of the nineteenth century the efforts scientifically to categorize women as naturally inferior to men intensified (and became more immediate) in the context of increasingly powerful threats to this presumption from women wanting to do 'male' things.[68] Women were also increasingly involved in campaigns against the exercise of medical authority, as over the issue of vivisection, or over the compulsory medical inspection of prostitutes.[69] In this context, at the same time that phrenology gave way to more detailed – if ultimately no less ambiguous – anthropometric studies of gender (as well as of age, race and class), the male scientific search for the essence of womanhood came to concentrate all the more intently on the female reproductive organs. By the time of Martineau's death, gynaecology had not only come to be identified with and defined as 'the science of womankind', and to be closely associated with Ploss-like natural histories of woman, but it had also rendered the boundaries between male and female in nature supposedly less tenuous through the identification and celebration of the ovaries as the quintessence of femininity – the 'soul of woman'. Although the ovaries were to be seen as equivalent to testes in men,[70] there was no return to the view of women as inverted men (which would have reestablished homologies and rendered men as open to biological reduction as women). The focus on the ovaries – whatever else this may have accomplished – extended the biological reduction of women and further rendered their interiors amenable only to the comprehensions of gynaecological experts.

At the same time, in a move that can hardly be regarded as separable from the social threat to this achievement from the women's movement, political efforts were under way within the emergent specialism of gynaecology to consolidate its area of expertise by better defining its own professional boundaries.[71] Internally and externally, therefore, gynaecology might be seen in much the same position that medicine as a whole had been when the Martineau case first emerged, with concerns over status and boundary definition being uppermost.

CONTRADICTIONS

In this altered context and for this professionalizing group in particular, the revelations of Martineau's post-mortem might be projected as a cause for jubilation. Martineau, after all, had been one of the most visible advocates for what was now to be seen, biologically, as the wrong-headed political struggle for women's equality as based upon the old liberal theory of natural rights. Indeed, since her mesmeric cure, Martineau had become all the more involved in this struggle. In 1849 she became the secretary of Bedford College for Women, and in her regular contributions to the *Daily News* (after 1853) she wrote repeatedly in defence of women's interests. Besides protesting against the Contagious Diseases Acts and for reform of the laws on women's inheritance of property, she advocated a larger place for women in civic and professional life – including, not least, in medicine. Women midwives, paediatricians and gynaecologists especially, she believed, would be the 'best advocates' for women's equality.[72]

Thus to have evidence that Martineau's perception of her cure by mesmerism was mistaken – that she had in fact been dominated all her life not merely by her *uterus* as Greenhow and Clarke had suggested in the 1840s, but by her diseased *ovaries*[73] – was of great symbolic significance. Through the pathological identification of her femininity, the nation's most 'man-like woman' ceased to be man-like, and hence ceased to pose any threat to the 'science of womankind' in its effort to sharpen the boundaries between males and females. Allayed were social anxieties over women doing 'male' things; for once reduced to her ovaries, Martineau ceased to be other than merely another misguided female arguing against 'nature' for women's equality. The mesmeric 'cure', far from continuing to call medical understandings into question, now served instead to call seriously into question the validity of all women's claims to knowing.

Yet it may be also that we should regard the extended coverage on Martineau's post-mortem in the medical press, not as reflecting any sense of triumph over the enemy without (or at least not only this), but rather – given how the coverage of her case was confined wholly to pathological findings – as covert acknowledgement of the need to silence deeper sets of contradictions raised by and signified through her endorsement of mesmerism. Despite the pathological reduction, Martineau remained in some ways as problematic for the 'scientists of womankind' as she had been to the medical reformers of Wakley and Greenhow's generation.

The least of the problem was that she was clearly not against orthodox medicine as such, nor against medical expertise. While she cultivated mesmeric healing and turned increasingly to hydropathy and homoeopathy in her own battle for good health,[74] she not only pressed for women's entry into orthodox medicine, but had no hesitation in consulting the heart specialists Latham and Watson. She also drew up a will in which she left her body and her brain to medical research.[75]

More important is that Martineau, subsequent to her mesmeric cure, became a major popularizer of the positivist epistemology upon which scientific medicine had come to rest. In 1853 she offered the public the first English translation of Comte's *Positive Philosophy*, with its legitimation of the 'progressive' scientific materialist world view over previous supposedly metaphysical and theological dominations.[76] Satisfied with endorsing this philosophy, Martineau seems not to have been bothered politically by Comte's views on the biological differences between the sexes. Less surprising (given the great speed with which she apparently carried out the translation) is that she did not make the connections between positivism and her previous advocacy of bourgeois liberal political economy.[77] Such connections cannot be gone into here; nor can we do more than note that they were contested by some Owenite socialists in the 1840s partly on the basis of mesmerism as perceived vitalistically. Suffice to say that for these Owenites mesmerism was a means to positing an alternative to what was seen as the mechanical, sterile, externalizing and one-dimensional positivist metaphysic of bourgeois capitalism, with its 'irrational' dichotomies between subject/object, mind/body, male/female and capital/labour.[78] This needs mentioning, not because the socialist critique of the positivist world view was particularly influential at mid-century, or even understood, nor, least of all, because this critique articulated in any way with Martineau's mesmerism (as we shall see in a moment). It deserves mention, rather, because many of these same criticisms of the cramping nature of scientific materialist culture resurfaced in the last third of the nineteenth century in the spiritualist movement. Although it is a measure of the success of positivism that within the spiritualist movement issues of gender were disengaged from those of capital (thus leaving only a comparatively small space for politics), for the women who became involved with spiritualism it provided a means (as mesmerism had for Martineau) temporarily to invert dominant and prescriptive views of passive femininity as constructed around illness, silence and functional (de-eroticized) sexuality.[79]

Despite the fact that mesmeric experiences and spiritualistic ones were often sociologically and psychologically indistinguishable in practice, and that the enthusiasts of both types of experience (who were often one and the same) tended to regard the medical profession with deep suspicion, Martineau shared none of her contemporaries' passion for spiritualism. She positivistically abhorred it. In an article on witchcraft that she wrote for the *Edinburgh Review* in 1868, she attacked spiritualists, claiming that they make 'an objective world of their own subjective experience', and contrasted this with the work of physiological psychologists, who were 'proceeding, by observation and experiment, to penetrate more and more secrets of our intellectual and moral life.'[80] The latter, in other words, were acting like good gynaecologists of the mind. In so saying, Martineau was sharing an outlook identical to that of the 1845 reviewer of her *Letters on Mesmerism* in the *Edinburgh Medical and Surgical Journal* who, besides struggling to identify Martineau as hysterical, sought to liken mesmerism to the 'absurd and fanciful delusions' of primitive religious peoples, and juxtapose these to medical science's reasoning by positive 'facts'.

Thus, just as Martineau's identification with bourgeois political economy and liberalism had compromised the medical reformers' attack on her mesmerism in the 1840s, so, in the 1870s, her identification with and popularization of scientific materialism in the face of spiritualism seriously compromised any retrospectively delivered attack on her mesmerism by the new medical exponents of scientific progressivism. The scope for the attack on her mesmerism on the basis of the pathological revelations of her post-mortem was, in the short term at least, cancelled out by Martineau's endorsement of scientific medicine's dominant epistemology.

The paradox in all this, from medicine's point of view, was that Martineau had arrived at her positivist stance not in spite of her mesmeric experience, but in large part because of it. The 'new self' to emerge out of that experience was one that shed Christian theology for a wholly secular, naturalistic and materialistic account of reality. She now looked back on her sick-room writings as displaying weakness and self-pity.[81] Although some defenders of Christianity accused Martineau of dabbling in diabolical sorcery and supernaturalism through mesmerism, nothing was further from the truth.[82] For Martineau, mesmerism was increasingly to be seen as a convincing demonstration of the power of natural forces, although she believed that the natural laws governing the phenomenon had yet to be explained. In the *Letters on the Laws of Man's Nature and Development*, which she and Henry Atkinson published in 1851, mesmerism and the physiological psychology of phrenology were to serve as the basis for a socially shocking reduction of mind to matter.[83]

Compounding this paradox was the fact that hand in hand with Martineau's endorsement of the epistemology inherent in the 'science of womankind' went her attack on the construction and maintenance of patriarchal social relations. Stemming from her mesmeric experience was not only the perception that religion was at variance with the facts of material science, but also, 'that men have ever constructed the image of a Ruler of the Universe out of their own minds.'[84] For a mid-Victorian woman to argue this was to act far more counter-culturally than merely to advocate mesmerism. Although by then Christian theology was no longer being deployed for the subordination of females on the grounds of women's potential carnality, it had been widely resorted to by antifeminists in the 1840s as a means to locate woman in a 'pure and exalted state'.[85] Martineau, by rejecting Christianity, was toppling that pedestal image of women, and undermining the ideology of separate spheres for which it stood. It needs adding, though, that from the perspective of women's rights, Martineau's theological attack on separate spheres was increasingly pointless and anachronistic by the 1870s. By then, in the face of the far more powerful restrictions of biology and pathology, the window of opportunity for women that the attack on theology had afforded in the 1840s was, for most women, shut tight and sealed.

Hence to the final irony: the striking alignment of Martineau's rationalistic defence of mesmerism with the feminist *denunciation* of mesmerism by Mary Wollstonecraft in her *Vindication of the Rights of Women* (1792). For Wollstonecraft, mesmerism was nothing but the charlatanry of 'cunning men' and 'lurking leeches', who exploited the credulity of ignorant women who had not risen above 'vulgar prejudices' and 'irrational devotion'. To Wollstonecraft it seemed that mesmerists, by claiming to 'put to flight, as it is said, disorders that have baffled the powers of medicine', operated against 'the light of reason'. The 'whole tribe of magnetisers', she thought – the 'men who laid claim to this privilege, out of the order of nature' and who, 'by hocus pocus tricks, pretend to work a miracle' – were as despicable as religious priests of old in their exploitation of women. These were the new 'priests of quackery', though it is true they have not the convenient expedient of selling masses for souls in purgatory, or churches where they can display crutches, and models of limbs made sound by a touch or a word.'[86] What was from Wollstonecraft's perspective oppressive to women, was from Martineau's perspective liberationist, *vis-à-vis* both the traditional priesthood and the new one of male-dominated orthodox medicine and biology.

CONCLUSION

Whether all or some of these paradoxes were apparent to those in the medical profession in the late 1870s who took particular interest in Martineau's morbid pathology we are unlikely ever to know. Nor is it of great importance that we find out. Like the focus on Martineau's mesmeric cure itself, the speculation entered into here, as to the possible deeper significance of the gynaecological interest in her case, is important mostly for what it reveals about the construction of gender in Victorian medicine, and, in turn, about historical constructions of women's resistance to and complicity with dominant patriarchal relations.

While the case of Harriet Martineau reveals the dialectical construction of gender in medicine, simultaneously it exposes the limits of one-dimensional accounts of women's experience of oppression. Depicted in part is not merely how the dichotomies – professional/lay, orthodox/heterodox, nature/nurture, rational/irrational, external/internal, fact/value, material/spiritual, and, above all, male/female – were socially and ideologically negotiated at particular times and places according to specific needs and interests, but also, more importantly, how these dichotomies, in reality, held little water in and of themselves. It was, rather, in the fact that contemporaries – including Martineau herself – came to conceptualize reality in these oppositional terms that female oppression was legitimated. Martineau, to the extent that she experienced that oppression (as symbolized in the medical denial of her mesmeric cure), was able also to resist it. Throughout her life she continued to believe that women could be equal to men if only they followed her example and struggled hard enough. She did not grant that for most women such struggle was a luxury. Nor was she in a position to realize that, even as she voiced women's rights, the political space for resistance was being undermined through the bio-pathological identification of women as inherently diseased – that henceforth, women's insides would deny their political outsides. Despite the fact that the project of gynaecology was played out on her own body, Martineau did not perceive that the social and ideological had come to be concealed in the biological and that, thereafter, the struggle against patriarchal authority would require resisting biology. Though hardly unique in this respect, Martineau's life story testifies remarkably well to how the ground had shifted between the 1840s and 1870s. Her continued commitment to mesmerism, in particular, illuminates how once-potential venues for women's resistance to patriarchy in medicine were rendered otiose. Historically, however, as a means of assessing the changing nature of gender in medicine, the story of Harriet Martineau's engagement with mesmerism remains of value.

ACKNOWLEDGEMENTS

This paper was drafted and first delivered in 1984. The research was made possible through the generous support of the Wellcome Trust.

Helpful comments and criticisms were gained from audiences in Lancaster, Colchester, Manchester and London. For their detailed comments and encouragements, I am lastingly indebted to Marina Benjamin, Mary Ann Elston, Mary Fissell, Ruth Harris, Ludmilla Jordanova, Irvine Loudon, Ornella Moscucci, Tina Posner and Mike Shortland.

NOTES

1 'Women's history and men's history', editorial, *History Workshop Journal*, 19 (1985), p. 1.
2 See, respectively: Ellen Pollak, 'Feminism and the New Historicism: a tale of difference or the same old story?' *The Eighteenth Century*, 29 (1988), pp. 281–6; Claudine Herzlich and Janine Pierret, *Illness and Self in Society*, tr. E. Forster (Johns Hopkins University Press, Baltimore, 1987); Roy Porter and Dorothy Porter, *In Sickness and in Health: the British Experience 1650–1850* (Fourth Estate, London, 1988); Julia L. Epstein, 'Writing the unspeakable: Fanny Burney's mastectomy and the fictive body', *Representations*, 16 (1986), pp. 131–66.
3 For some of the best recent work, see Catherine Gallagher and Thomas Laqueur (eds), *The Making of the Modern Body: Sexuality and Society in the Nineteenth Century* (University of California Press, Berkeley, 1987); Ludmilla Jordanova, *Sexual Visions: Images of Gender in Science and Medicine between the Eighteenth and Twentieth Centuries* (Harvester/Wheatsheaf, Hemel Hempstead, 1989); Dorinda Outram, *The Body and the French Revolution: Sex, Class and Political Culture* (Yale University Press, New Haven and London, 1989); and the splendid collection of papers in M. Feher, Ramona Naddaff and Nadia Tazi (eds), *Fragments for a History of the Human Body, Zone*, vols 3, 4, 5 (New York, 1989).
4 Margaret Walters, 'The rights and wrongs of women: Mary Wollstonecraft, Harriet Martineau, Simone de Beauvoir' in Juliet Mitchell and Ann Oakley (eds), *The Rights and Wrongs of Women*, (Penguin, Harmondsworth, 1976), pp. 304–78. Martineau's rediscovery by twentieth-century feminists was signalled by Alice Rossi's inclusion of Martineau's *Society in America* (1837) in her edition of *The Feminist Papers* (1973; repr. Northeastern University Press, Boston, 1988). Other feminist writings on Martineau, not mentioned below, include Gayle Graham Yates (ed.), *Harriet Martineau on Women* (Rutgers, New Brunswick, N.J., 1985); Gaby Weiner, 'Harriet Martineau: a reassessment' in Dale Spender (ed.), *Feminist Thinkers: Three Centuries of Key Women Thinkers* (Pantheon, New York, 1983), pp. 60–74; and Mitzi

Meyers, 'Harriet Martineau's *Autobiography*: the making of a female philosopher' in Estelle C. Jelinek (ed.), *Women's Autobiographies* (Indiana University Press, Bloomington, 1980), pp. 53–70.

5 William Howitt, 'Harriet Martineau', *The People's Journal*, 14 March 1846, p. 141. See also R. K. Webb, *Harriet Martineau: A Radical Victorian* (Heinemann, London, 1960), esp. pp. 140–1; V. K. Pichanick, *Harriet Martineau: The Woman and her Work, 1802–76* (University of Michigan Press, Ann Arbor, 1980) and Valerie Sanders, *Reason over Passion: Harriet Martineau and the Victorial Novel* (Harvester Press, Brighton, 1986), ch. 7, '"The most manlike woman in the three kingdoms": Harriet Martineau and feminism'.

6 Deirdre David, *Intellectual Women and Victorian Patriarchy: Harriet Martineau, Elizabeth Barrett Browning, George Eliot* (Macmillan, London, 1987), pp. viii–xii.

7 Cf. Sylvana Tomaselli, 'Collecting women: the female in scientific biography', *Science as Culture*, 4 (1988), pp. 95–106, esp. pp. 105–6; Maureen McNeil, 'Being reasonable feminists' in Maureen McNeil (ed.), *Gender and Expertise* (Free Association Books, London, 1987), pp. 13–61; and sources at n. 3 above.

8 T. M. Greenhow, *Medical Report of the Case of Miss H —— M ——* (Samuel Highley, London, 1845), pp. 10–11.

9 Ibid., pp. 13–16.

10 See her letters covering this period in *Harriet Martineau's Letters to Fanny Wedgwood*, ed. E. S. Arbuckle (Stanford University Press, Stanford, 1983), esp. p. 22. In this respect, as in so many others, Martineau was like Florence Nightingale.

11 Published anonymously, *Life in the Sick Room: Essays by an Invalid* (Edward Moxen, London, 1844) ran to a 3rd edn in 1849. Chapters on euthanasia were omitted by the publisher: see Mrs F. Fenwick Miller, *Harriet Martineau* (W. H. Allen, London, 1884), p. 123. For discussion of the text, see Porter and Porter, *Sickness and Health*, pp. 212–17; and Webb, *Martineau*, pp. 193 ff.

12 Diana Postlethwaite, 'Mothering and mesmerism in the life of Harriet Martineau', *Signs*, 14 (1989), pp. 583–609 at pp. 589–90, who draws on Mitzi Meyers, 'Unmothered daughter and radical reformer: Harriet Martineau' in Cathy N. Davidson (ed.), *The Lost Tradition: Mothers and Daughters in Literature* (F. Ungar, New York, 1980), pp. 70–9.

13 *Life in the Sick Room*, p. xi.

14 See Greenhow, *Medical Report*, p. 18; Martineau, *Letters on Mesmerism* (Edward Moxen, London, 1845), p. 5; Martineau, *Autobiography* (3rd edn, Smith, Elder & Co., London), vol. 2, p. 191; S. T. Hall, *Mesmeric Experiences* (H. Bailliere, London, 1845), pp. 63–75. The predisposition to

experiment with mesmerism is suggested in her comments on mesmerism in *Life in the Sick Room*, pp. 82–3. Besides referring to mesmerism as 'powerful', and 'occult and mysterious', she also referred to it as 'this beautiful natural remedy': see *Letters to Wedgwood*, p. 79.

15 Fred Kaplan, ' "The mesmeric mania": the early Victorians and animal magnetism', *Journal of the History of Ideas*, 35 (1974), pp. 691–702. On Hall and others involved with phreno-mesmerism, see Roger Cooter, *Phrenology in the British Isles: An Annotated, Historical Biobibliography and Index* (Scarecrow Press, Metuchen, NJ, 1989).

16 'On the absurdities of mesmerism', *London Medical Gazette*, (1843/4), p. 705.

17 See *John Elliotson on Mesmerism*, ed. Fred Kaplan (De Capo, New York, 1982); Jonathan Miller, 'A Gower Street scandal', *Journal of the Royal College of Physicians*, 17 (1983), pp. 181–91; and T. M. Parssinen, 'Professional deviants and the history of medicine: medical mesmerists in Victorian Britain' in Roy Wallis (ed.), *On the Margins of Science: the Social Construction of Rejected Knowledge* (University of Keele, 1979), pp. 103–20.

18 J. Pearson, *A Plain and Rational Account of Animal Magnetism* (London, 1790), pp. 4, 18.

19 However, the introduction of chloroform in 1846 was to raise new sets of gender problems; see Mary Poovey, ' "Scenes of an Indelicate Character": the medical "treatment" of Victorian women' in Gallagher and Laqueur (eds), *Modern Body*, pp. 137–68.

20 *Lancet*, 1 Feb. 1845, p. 140.

21 Martineau, letter to Wedgwood, 19 Mar. 1845 in *Letters to Wedgwood*, p. 79.

22 'Mesmerism, Miss Martineau, and the "Great New Idea" ', *Lancet*, 30 Nov. 1844, p. 291.

23 Elliotson, 'Miss Martineau and her traducers', *Zoist*, 3 (1845), p. 86.

24 Martineau, letter to H. G. Atkinson, 18 Sept. 1874, quoted in Miller, *Martineau*, p. 131.

25 'Pretensions of mesmerism as a therapeutic agent', *Edinburgh Medical and Surgical Journal*, 63 (1845), pp. 464–96, at p. 479.

26 'The hysterical woman' in Carroll Smith-Rosenberg, *Disorderly Conduct: Visions of Gender in Victorian America* (Knopf, New York, 1985), pp. 197–216. For further sources and discussion, see Postlethwaite, 'Martineau', pp. 589–90. On female illness generally as a way of resisting medical definition, see Jeffrey Weeks, *Sex, Politics and Society: The Regulation of Sexuality since 1800* (Longman, London, 1981), p. 45.

27 Quoted in *Lancet*, 15 Feb. 1845, p. 213.

28 Charles Radclyffe Hall, 'On the rise, progress, and mysteries of mesmerism in all ages and countries', *Lancet*, Feb.–May 1845. On Martineau and mesmerism, see 12 Apr. 1845, p. 403.

29 Greenhow, *Medical Report*, pp. v, 17.

30 Hall, letter to *Atlas*, 27 Jan. 1845, reprinted in his *Mesmeric Experiences*, pp. 63–75; and Elliotson, 'Martineau and her traducers', pp. 92 ff.

31 See Roger Cooter, *The Cultural Meaning of Popular Science: Phrenology and the Organization of Consent in Nineteenth Century Britain* (Cambridge University Press, Cambridge, 1984), pp. 32 ff and *Phrenology in the British Isles*.

32 *Society in America*, 2nd edn (Saunders and Otley, 1837), vol. 3, pp. 103, 150–1, and vol. 1, pp. 200–1.

33 [Mary Astell], *An Essay in Defence of the Female Sex*, 3rd edn, (London, 1696), p. 21; see also Francis Power Cobbe, 'Criminals, idiots, women and minors: is the classification sound?' *Fraser's Magazine*, 78 (Dec. 1868), p. 778; and John S. Haller and Robin M. Haller, *The Physician and Sexuality in Victorian America* (University of Illinois Press, Urbana, 1974), p. 52.

34 [A. W. Kinglake], 'The rights of women', *Quarterly Review*, 75 (1844/5), pp. 94–125 at p. 111.

35 [T. H. Lister], 'Rights and conditions of women', *Edinburgh Review*, 73 (1841), pp. 189–209. For the feminist response, see Mrs Hugo Reid, *A Plea for Women* (W. Tait, Edinburgh, 1843).

36 See Cooter, *Cultural Meaning*, pp. 368–9 and 'The history of mesmerism in Britain: poverty and promise' in Heinz Schott (ed.), *Franz Anton Mesmer und die Geschichte des Mesmerismus* (Steiner, Stuttgart, 1985), pp. 153–62; and Barbara Taylor, *Eve and the New Jerusalem: Socialism and Feminism in the Nineteenth Century* (Virago, London, 1983).

37 *The Critic*, March 1844, p. 108.

38 At least three such cases of mesmeric referral (among private patients) were reported by the obstetric physician Robert Lee in 1855 – a further sixteen such women having also had recourse to hydropathic treatment and/or homoeopathic treatment: Lee, 'The diagnosis and treatment of uterine diseases', *Medico-Chirurgical Transactions*, 38 (1855), pp. 289–342, esp. pp. 304, 317, 318. All three of the women who had consulted mesmerists were unmarried.

39 On hypnosis, especially in relation to the treatment of hysteria, see Elaine Showalter, 'Feminism and hysteria: the daughter's disease' in Showalter, *The Female Malady: Women, Madness, and English Culture, 1830–1980* (Pantheon, New York, 1985), pp. 145–64; and Ruth Harris, *Murders and Madness: Medicine, Law and Society in Fin de Siècle* (Clarendon Press, Oxford, 1989).

40 T. M. Greenhow, 'Termination of the case of Miss Harriet Martineau', *British Medical Journal*, 14 Apr. 1877, pp. 449–50. On Martineau's fear of this, see her letter to H. G. Atkinson, 18 Sept. 1874, quoted in Miller, *Martineau*, p. 131.

41 The post-mortem, by W. Moore King, is quoted by Greenhow, 'Termination', pp. 449–50.

42 T. Spencer Wells, 'Remarks on the case of Miss Martineau', *British Medical Journal*, 5 May 1877, pp. 543, 550, 647–8. The cyst was presented to the Clinical Society by Martineau's other medical brother-in-law, Alfred Higginson of Liverpool. Wells's speech to the Society was preceded by a reading of the post-mortem report.

43 W. O. Markham, 'The case of Miss Martineau', *British Medical Journal*, 17 Nov. 1877, p. 712.

44 Carlyle, quoted in Pichanick, *Martineau*, p. 133; Martineau, *Autobiography*, vol. 2, p. 201.

45 Webb, *Martineau*, p. 20. For Atkinson's part in her treatment, see Martineau, *Autobiography*, vol. 2, pp. 213–15.

46 See *Letters on Mesmerism*; *Letters to Wedgwood*, pp. 79–92, 106; Miller, *Martineau*, p. 147; and John Elliotson, 'Mesmeric cure of a cow, by Miss Harriet Martineau', *Zoist*, 8 (1850), pp. 300–3.

47 'Mesmerism revived', *British Medical Journal*, 29 July 1876, p. 161, and ibid. 1 July 1876, p. 20.

48 *Autobiography*, vol. 2, p. 191.

49 After Latham examined her on 23 Jan. 1855 and tentatively diagnosed an enlarged heart, Sir Thomas Watson was consulted. He suggested that her tumour had only migrated at the time of the supposed mesmeric cure. Though Martineau confided to John Chapman after Feb. 1855 that she still had the tumour, she declared publicly that she had a heart complaint: *Letters to Wedgwood*, p. 131. See her letter of 16 Sept. 1855, written in the 'strictest confidence' to 'Dear Friend' (?Atkinson), acknowledging 'a very large internal tumour'. Bodleian Library, Oxford, Eng. Lett d2, fols 185–9. Cf. letters quoted in Miller, *Martineau*, pp. 130–2.

50 The *Oxford English Dictionary* attributes 'gynaecology' to John Craig, 1847, but a decade before this Michael Ryan, in his *The Philosophy of Marriage . . . with an account of the diseases of the genito-urinary organs* (Churchill, London, 1837), wrote that 'The science of anthropology or andrology, gynaecology and paedology, is one of the most important and positive in medicine', and referred to 'Gynaecopathology' as embracing the diseases peculiar to women (pp. 19–20).

51 'Modes of obtaining "Eminence" in the profession', *Lancet*, 1 Aug. 1846, pp. 133–4. See also *British and Foreign Medico-Chirurgical Review*, 15 (Apr. 1855), p. 315, and Poovey, 'Indelicate Character', pp. 144–5.

52 The physician Marshall Hall, shortly after moving to London and publishing *Commentaries on Some of the More Important of the Diseases of Females* (Longman, London, 1827), was earning £800 p.a.; by 1833 this had

risen to £2,200. See Charlotte Hall, *Memoirs of Marshall Hall* (R. Bentley, London, 1861), pp. 69–70.

53 E.g. M. Hall, *Diseases of Women*, p. x; Ryan, *Philosophy of Marriage*, p. 1; and John Roberton, *Essays and Notes on the Physiology and Diseases of Women, and on Practical Midwifery* (Churchill, London, 1851), p. 203.

54 E.g. Robert Ferguson, *Essays on the Most Important Diseases of Women* (John Murray, London, 1839); John Lever, *A Practical Treatise on Organic Diseases of the Uterus* (Longman, London, 1843); Samuel Ashwell, *Treatise on Diseases Peculiar to Women* (S. Highley, London, 1844); E. J. Tilt, *On the Preservation of the Health of Women at the Critical Periods of Life* (Churchill, London, 1851).

55 For the classic texts in this genre, see G. J. Barker-Benfield, *The Horrors of the Half-Known Life: Male Attitudes Toward Women and Sexuality in Nineteenth Century America* (Harper and Row, New York, 1976), and Barbara Ehrenreich and Deirdre English, *For Her Own Good: 150 Years of the Expert's Advice to Women* (Pluto Press, London, 1979). Most such work focuses on America and on the late nineteenth century.

56 E.g. Thomas Gisborne, *An Enquiry into the Duties of the Female Sex* (J. Davis, London, 1795).

57 Alexander claimed to be treading 'a path which ha[d] never been attempted before.' Although earlier works had been written in defence of female equality, as, for example, that by Mary Astell (n. 33 above), of which there were editions to 1750, these had not conceptualized woman as a category of natural historical inquiry. A partial exception was John Millar's *Origin of the Distinctions of Ranks* (London, 1771), on which see Thomas Laqueur, 'The politics of reproductive biology' in Laqueur and Gallagher (eds), *Modern Body*, pp. 21–2. Alexander's work reached a 3rd edn in 1782 and was translated into French and German.

58 The first volume of which was published in 1885, the third in 1927; an English translation, by Eric Dingwall, appeared in 1935; selections are given in Paula Weideger's *History's Mistress: A New Interpretation of a Nineteenth-Century Ethnographic Classic* (Penguin, Harmondsworth, 1986).

59 James Ricci, *One Hundred Years of Gynaecology, 1800–1900* (Blakiston, Philadelphia, 1945), p. 32.

60 See especially Laqueur, 'Reproductive biology', p. 24.

61 Quoted in Londa Schiebinger, 'Skeletons in the closet: the first illustrations of the female skeleton in eighteenth-century anatomy' in Gallagher and Laqueur (eds), *Modern Body*, pp. 42–82, at p. 69.

62 See Poovey, 'Indelicate character', p. 145.

63 Astell, *Female Sex*, p. 21; Alexander, *History of Woman*, p. 36.

64 *Observations on those Diseases of Females Which are Attended by Discharges* (Longman, London, 1814), p. 2.

65 See Henry G. Wright, *Uterine Disorders: Their Constitutional Influence and Treatment* (Churchill, London, 1867), p. 3.

66 Letter of 5 Oct. 1842, summarized in the typescript of letters (1819–1845) by W. S. Coloe, p. 173. Bodleian Library, Oxford.

67 Ricci, *Gynaecology*, p. 25.

68 Cynthia Russett, *Sexual Science: The Victorian Construction of Womanhood* (Harvard University Press, Cambridge, Mass., 1989).

69 See Brian Harrison, 'Women's health and the Women's Movement' in Charles Webster (ed.), *Biology, Medicine and Society, 1840–1940* (Cambridge University Press, Cambridge, 1981), pp. 15–71, esp. p. 24; Mary Ann Elston, 'Women and anti-vivisection in Victorian England' in Nicolas Rupke (ed.), *Vivisection in Historical Perspective* (Routledge, London, 1987), pp. 259–94; and Judith R. Walkowitz, *Prostitution and Victorian Society* (Cambridge University Press, Cambridge, 1980).

70 See Edward John Tilt, *On Diseases of Menstruation and Ovarian Inflammation, in Connexion with Sterility, Pelvic Tumours and Affections of the Womb* (Churchill, London, 1850), pp. xix, 7; cf. Laqueur, 'Reproductive biology', p. 2.

71 See Ornella Moscucci, *The Science of Woman: Gynaecology and Gender in England 1800–1929*, (Cambridge University Press, Cambridge, 1990).

72 *Autobiography*, vol. 1, pp. 400 ff. For a list of her contributions to the *Daily News*, see the handlist compiled by R. K. Webb, Bodleian microfilm reel 1216.

73 Greenhow and Clarke never referred to Martineau's ovaries, nor to the possibility of an ovarian cyst, despite the fact that knowledge of such cysts and their removal by surgery was highly controversial by the 1840s. Seven successful cases of removal were reported in 1843 alone: see Ricci, *Gynaecology*, pp. 48, 64–5.

74 See her letters of 1856 and 1857 in Bodleian Library: Eng. lett. d2 fols 190, 205; *Life in the Sick Room*, pp. 83–4; and Webb, *Martineau*, p. 248.

75 Initially her body was willed to the Unitarian Southwood Smith (who had dissected Bentham); later it was willed to William Moore King. King betrayed Martineau's trust in him by allowing Higginson to appropriate the cyst (see nn. 40 and 42 above). The brain was to go to Atkinson, who was a phrenologist. See *Letters to Wedgwood*, p. 139; and Miller, *Martineau*, pp. 174, 209.

76 *The Positive Philosophy of Auguste Comte* freely translated and condensed by Harriet Martineau, 2 vols (John Chapman, London, 1853; 2nd edn, 1875).

77 On positivism, as essentially the metaphysic of capitalism, and for further references, see Roger Cooter, 'Anticontagionism and history's medical record' in P. Wright and A. Treacher (eds), *The Problem of Medical Knowledge: Examining the Social Construction of Medicine* (Edinburgh University Press, Edinburgh, 1982), pp. 87–108, esp. p. 91 and 'The conservatism of

"Pseudoscience"' in Patrick Grim (ed.), *Philosophy of Science and the Occult* (State University of New York Press, Albany, 1982), pp. 130–43.

78 See Cooter, 'Mesmerism in Britain'; and Taylor, *Eve*.

79 Alex Owen, 'Women and nineteenth-century spiritualism: strategies in the subversion of feminity' in Jim Obelkevich, L. Roper and R. Samuel (eds), *Disciplines of Faith: Studies in Religion, Politics and Patriarchy* (Routledge, London, 1987), pp. 103–53.

80 'Salem witchcraft', *Edinburgh Review*, 128 (July 1868), pp. 1–47, at pp. 1, 47. Forgetting that she had written on the subject in 1834 from a Christian point of view, she regarded this article as 'terribly important': see Webb, *Martineau*, p. 249. Cf. W. B. Carpenter, 'Mesmerism, odylism, table-turning, and spiritualism considered historically and scientifically', *Fraser's Magazine*, 95 (Feb. 1877), pp. 135–57. Martineau never referred to achievements in gynaecology.

81 *Autobiography*, vol. 2, pp. 169–74.

82 See Charlotte Elizabeth Tonna (editor of *Christian Ladies Magazine*), *Mesmerism: A Letter to Miss Martineau* (Seeley, Burnside and Seeley, London, 1844), cited in Vera Wheatley, *The Life and Work of Harriet Martineau* (Secker and Warburg, London, 1957), p. 239. However, Martineau once experimented with table rapping: see her letter to John Chapman, 3 Dec. 1857 in Eng. lett, d2 fols 207–8, Bodleian Library.

83 The strongest reaction was from her brother James: 'Mesmeric atheism', *Prospective Review*, 7 (Apr. 1851), pp. 224–62. See also Dr J. S. Bushnan, *Miss Martineau and her Master* (Churchill, London, 1851).

84 Miller, *Martineau*, p. 152, referring to the revelations of Martineau's *Eastern Life, Past and Present* (Edward Moxen, London, 1848).

85 See, for example, 'Woman and social system', *Fraser's Magazine*, 21 (1840), pp. 689–702 at p. 695. On Christianity and women's carnality, see L. Davidoff, 'Class and gender in Victorian England' in Judith L. Newton, M. P. Ryan and J. R. Walkowitz (eds), *Sex and Class in Women's History* (Routledge, London, 1983), pp. 17–71, at pp. 21 ff.

86 Norton Critical Edition, ed. Carol H. Poston (New York and London, 1975), pp. 179–81.

6

Hermaphroditism and Sex Difference: The Construction of Gender in Victorian England

Ornella Moscucci

Man with the head, and woman with the heart;
Man to command, and woman to obey;
All else confusion.

Tennyson, *The Princess* (1847)

During the Victorian period the value of science was widely proclaimed. Manufacturers and merchants were among the first to realize that science could bring practical benefits to industry. This prompted the establishment of new scientific institutions, for example the University of London, and led to the reform of ancient educational establishments such as Oxford and Cambridge. But the final appeal of science went far beyond its pecuniary advantages. Science appeared to many to be the new road to certainty, the only means of obtaining absolute truths about humankind and the universe. The period with which we are concerned saw the development of positivism, not only as scientific method, but also as philosophical doctrine and social theory. In essence, positivism was an assertion of faith in the growth of objective knowledge and in the value of science as a tool for social progress. At a time of increasing scepticism about the interpretative power of metaphysics, conservatives and reformers alike looked to science as the new foundation for political and social action. The most controversial issues of the day, from the emancipation of blacks to the Irish Question, were to be submitted 'to Agassiz and Huxley, not to Kant or Calvin, church or Pope';[1] comparative anatomists, biologists and physiologists had to be the arbiters of social change, because the status and social role of individuals derived from the just appreciation of their physical peculiarities and 'natural' aptitudes.

Bio-medical writers were especially confident that science could make a vital contribution to the 'Woman Question', the debate over women's rights and social responsibilities,[2] for they widely believed that the characters of sex were more evident than any other physical attribute displayed by human beings. As the Scottish-trained gynaecologist James Jamieson (1840–1916) wrote in 1887, there were various taxonomic devices round which humankind could be grouped – for example, the comparative degree of civilization of different races, their colour, or their speech. There was, however, 'one obvious division, more fixed and definite than any of these, and that is according to sex.'[3] Sexual classification involved drawing precise demarcation lines between men and women, which hinged on oppositions and contrasts between key anatomical and physiological properties. Laqueur has cogently argued that this emphasis on the incommensurability of male and female bodies constituted a radical departure from ancient and Renaissance accounts of masculinity and femininity.[4] Since the times of Galen, medical writers had posited an exact physiological homology between male and female reproductive organs, the chief difference between the sexes being the amount of 'heat' peculiar to each. By the late eighteenth century this model was in retreat and scientific representations of gender began to emphasize the very different import of reproduction in the economy of the female as compared with the male's. Bound to their physical nature, women became a race apart by virtue of their reproductive functions, whereas men, who were not so restricted by their biology, were potential members of the broadest social groups.

The ideological status of the male/female dichotomy is now well documented; so is its role in underpinning the separation of the public and the private, a social and political distinction which has been central to the legitimation of Western democracy. But this is just one aspect of the conceptual possibilities engendered by the use of dichotomies, for these do not function only at the level of difference. As Jordanova has observed, by pairing separate terms together the idea of a *kinship* between them is evoked.[5] Jordanova considers the mind/body dichotomy and the new concepts which were created during the eighteenth century to bridge the gap between its constituent terms: notions such as 'sensibility', 'habit' and 'organization' referred to *both* mental *and* bodily processes, giving rise to productive ambiguity and a vast range of potential uses. Developing this point in relation to the man/woman opposition, we should ask whether bio-medical writers did recognize the existence of a similarity between the masculine and the feminine and, if so, what was the nature of the kinship?

Such a bridging concept may be readily found in ideas about the latent hermaphroditism of man and woman. This notion was important not only because it served to throw light on each of the paired terms, but also because it enabled bio-medical writers to explore the meaning of a human nature common to both sexes. The fact that the fundamental bisexuality of the human species was conceived of as a dormant property is crucial: the concept of latency created the possibility that human beings might fulfil their androgynous potential, raising questions about the natural or environmental conditions which could lead to the blurring of sex differences. The notion of latent hermaphroditism thus provided an invaluable tool with which such questions as the interaction between nature and culture and the historicity of human nature could be explored. I shall illustrate these issues by drawing on nineteenth-century medical and anthropological texts.

THE TAXONOMY OF SEX

During the late eighteenth century the application of scientific techniques to every aspect of human life was widely advocated. At a time when political tyranny and the power of the clergy were being criticized, many writers envisaged this study as the source of new moral precepts by which people should live in a reformed society. Nature had to provide the means of unmasking the artificiality and injustice of existing social practices such as forced marriages, inheritance laws, wars and the Inquisition; once the natural laws which governed human life had been discovered, society could be reorganized on a rational basis, independently of the arbitrary will of individuals, and human happiness secured.[6]

The science of Man took many forms, from the analysis of mental operations to descriptive studies of behaviour, custom and law. The third major component of the 'science of Man', and the one with which we are chiefly concerned in this chapter, was a classificatory science of the human species, or natural history of Man. Central to this study was the application of comparative anatomy and physiology to the analysis of human diversity: starting from the premise that the human species should be treated just like any other animal species, bio-medical investigators compared and contrasted the physical varieties of humankind in order to determine their place in the scheme of the universe. A keen interest in the male and female forms of Man was invariably present in the work of Enlightenment bio-medical writers. Reliable criteria of sexual classification were sought not only for the purpose of identifying which characters were sex-specific and which were peculiar to the human species, but also with a view to establishing the 'natural' order of the sexes in relation to each other. As democracy challenged the legitimacy of patriarchal authority, it was a matter of the greatest urgency that sex roles should be re-examined in the light of new 'scientific' principles.

Increasingly during the late eighteenth century, the biological difference between man and woman was emphasized. As the French physician and *idéologue* Cabanis wrote, 'nature has not simply distinguished the sexes by a single set of organs, the direct instruments of reproduction: between men and women there exist other differences of structure which relate more to the role which has been assigned to them.'[7] For Rousseau, men and women were demonstrably the same in everything that was not associated with sexual differentiation; in everything connected with sex they were related, but different. However, there was an important asymmetry in that the male transcended his sexual nature, whereas the female was female 'her whole life': childbearing, suckling and nurturing entirely dominated woman's organization.[8] This association of femininity with natural processes defined woman as a problem for scientific enquiry. It was during the Enlightenment that the study of feminine nature began as a genre combining scientific findings with literary observation and philosophical aphorisms.[9] From Pierre Roussel's *Système physique et moral de la femme* (1777) to William Alexander's *History of Women, from the Earliest Antiquity, to the Present Time* (1779), Enlightenment writers asserted the anomalous nature of woman in relation to man, the gold standard of anthropological discourse.[10]

However, there was a problem with this conception of sexuality, which Enlightenment writers bequeathed to their nineteenth-century heirs. To analyse men and women in terms of differences and contrasts was not entirely satisfactory from the anthropological point of view, for there was a danger of obscuring racial categories and undermining the unity of the human species in relation to the lower animals. This dilemma was clearly articulated by the mid-Victorian scientist George John Romanes (1848–94) in an essay entitled 'Mental differences between men and women', which first appeared in *Nineteenth Century* for May 1887: 'While within the limits of each species the male differs psychologically from the female,' Romanes wrote, 'in the animal kingdom as a whole the males admit of being classified, as it were, in one psychological species, and the females in another.' This was a bold statement, which Romanes hastened to correct. He did not mean to say 'that there is usually greater psychological difference between the two sexes of the same species than there is between the same sexes of different species.' he explained. 'I mean only that the points wherein the two sexes differ psychologically are more or less similar wherever these differences occur.'[11]

The nineteenth-century fascination, one might even say obsession, with the latent hermaphroditism of humankind can be seen as an attempt to reconcile the concept of sexual difference with the idea of a human nature common to both sexes. Hermaphroditism acted as a bridge between male and female – categories that were commonly defined in terms of oppositions and contrasts – forming an undifferentiated middle ground where the distinguishing characters of a species could be properly discerned. Many examples could be given to illustrate how this idea functioned in biology, physiology and embryology, but one that stands out with particular clarity is the nineteenth-century belief in male menstruation.

The notion that blood periodically issued from men as it did from women had a long history, which went back to Greek and Roman times. It can be understood in the light of the Greek conception of menstruation as a sort of physiological blood-letting by which women rid themselves of a superfluous quantity of blood. Whenever the function was in abeyance, periodic haemorrhages 'vicarious' of the menstrual flux would ensue from other parts of the body – such as the nose, the anus, the gums, the kidneys or the nipples – in order to re-establish the equilibrium of humours.[12] Given that menstruation was considered to be no different from other forms of bleeding, it was only a short step to interpret periodical haemorrhages in the male as cases of 'vicarious' menstruation.

A first-hand account of a 'case of menstruation in the male' was provided in 1867 by the American physician V. O. King. The menstruating male – a fellow student at the University of Louisiana medical school – 'periodically performed the simulated functions of menstruation', though 'not possessed of the usual organs'. When King first witnessed the phenomenon, the youth

had been the victim of this vicarious function for a period of three years, eliminating an apparent catamenial secretion, with the same regularity, and attended by the same indications by which it is characterised in the human female. The fluid exuded, flowed from the sebaceous glands of the deep fossa behind the corona glandis, and was of a sanguineous appearance, homogeneous and thick. The quantity of this exudation varied from one to two ounces during each haemorrhagic period, and the duration of the periods from three to six days.[13]

King had considered the possibility that the discharge might be caused by venereal disease, but he was emphatic that his friend had never suffered from such afflictions.

Similar cases were reported by the leading Victorian obstetrician, Alfred Wiltshire, in an article published in the *Lancet* for 1885. The French physician Chopart mentioned a young soldier who had a monthly discharge of bloody urine, accompanied by all the symptoms characteristic of menstruation; his fellow countryman Rayer cited the case of a butcher from Sedan, 'whose infirmity, becoming known, inspired so great disgust that no one would purchase meat from him.' Wiltshire himself knew of a surgeon who had bleeding from his penis every three weeks.[14]

Wiltshire took the view that the occurrence of vicarious haemorrhages in the male could be understood in terms of the theory of 'vital periodicity'. This doctrine had first been proposed in the early 1840s by the renowned English neurophysiologist Thomas Laycock (1812–76), who was also the author of a famous treatise on women's mental diseases. Laycock believed that all physiological phenomena were governed by regular temporal cycles, which he ascribed to the influence of the sun, the moon and the seasons. Vicarious haemorrhages and menstruation itself, he maintained, were nothing more than manifestations of this fundamental law of periodicity.[15] As the law affected all living beings, regardless of their sex, mere logic dictated that haemorrhages analogous to the menstrual flux could occur in men as well as women. 'All periodical haemorrhages in the human species', wrote Alfred Wiltshire, 'are under the dominion of the primal law of periodicity, which it inherits in common with all animals; only in the female we see, as a matter of observation, that the influence of periodicity is most markedly displayed.'[16]

The difference between male and female was thus not one of essence, but of degree. The Viennese laryngologist Wilhelm Fliess (1858–1928), one of Freud's closest collaborators, reached much the same conclusions when he set out to explain the nature of periodical nosebleeds in the male. This phenomenon suggested to him that two biological cycles, lasting twenty-eight and twenty-three days each, were at work in all physiological processes. Both periods were present in both sexes, he argued; but while the first was dominant in woman, the second was more marked in man. This 'discovery' led Fliess to formulate his theory of the essential bisexuality of human beings: 'These two groups of periodic processes . . . have a solid inner relation with male and female sexual characteristics. And it is only in accordance with our actual bisexual constitution if both – only with different stress – are present in every man and woman.'[17] Freud was very impressed with this conception of human physiology, and he later adopted it as the cornerstone of his theory of the aetiology of psychoneuroses.

It was not only at the psychological and physiological levels that the fundamental similarity between male and female was manifested. During the nineteenth century, theories of embryonic development re-proposed the ancient Galenic homologies within a new historical framework.[18] Physiologists believed that the penis and the clitoris, the scrotum and the labia, the testes and the ovaria, and so on, shared common origins in foetal life. During the period from conception to puberty they became differentiated in structure and function, but to each organ in the male, there corresponded a homologous structure in the female. Some organs became fully developed in the adults of one sex only, but could also be detected in the other sex in the embryonic state: for example, the Wolffian ducts developed into Fallopian tubes for the female, the cornua of the male utriculus into *vasa deferentia* for the male. Other organs, for example, the breasts, were functionally specific to the adult of one sex only, but were also present in the other in a rudimentary type of structure. The obstetrician Arthur Farre, for one, believed that the uterus itself was not 'altogether peculiar to the female', having 'its representative in the male, though only in a rudimental state.'[19]

Farre's words make the point that it was not only the gendering of individuals, but also that of the characters themselves which was problematic. A telling example is provided by Sir James Young Simpson (1811–70), who was the author of an important work on hermaphroditism published in 1839. The Edinburgh obstetrician considered whether the development of the breasts was a standard character of the human species or a feminine peculiarity. One would have expected the latter to be the case, Simpson argued, in view of woman's role in reproduction. However, well-developed breasts were occasionally found in hermaphrodites who were predominantly of the *male* type, and even masculine men sometimes secreted milk through their nipples. This seemed to show

MADEMOISELLE LEFORT.
Exhibited in Spring Gardens, 1818.
Published for R.S.Kirby, 11. Warwick Lane, Oct.ᵗ 1.1819.

PLATE 6 Mlle Lefort. Hermaphrodites frequently earned a living from exhibiting themselves in public.

(Reproduced by kind permission of the Wellcome Institute Library, London) that the development of the breasts was not the exclusive property of the female, but was common to both sexes and therefore proper to the species in general.[20]

Bound up with the problem of sexual differentiation was the question of whether the man or the woman was the standard of the species. The seeming impossibility of identifying unambiguous sex characters had persuaded Simpson that the only true representative of a species was the hermaphrodite: 'the natural characters of any species of animal', he wrote,

> are certainly not to be sought for solely either in the system of the male or in that of the female; but . . . they are both to be found in those properties which are common to both sexes, and which we have seen combined together by nature upon the bodies of an unnatural hermaphrodite, or evolved from the interference of art upon the castrated male or spayed female.[21]

Simpson believed that the secondary sexual differences depended on the ovaries in the female and the testes in the male. When the activity of these organs was in abeyance, for example before puberty or after the menopause, people closely approximated the type of the species. Of the two sexes, it was the male who at puberty departed from the androgynous state common to the young of both sexes. This marked the greater perfection of the male as an individual; however, the female was his superior from the point of view of the species, since her androgynous appearance more truthfully represented the characters of the human species.[22]

Evolutionist biology was to put paid to the idea that the hermaphrodite was the standard of the species. By the last quarter of the nineteenth century, hermaphroditism was seen less positively as a lower stage of development from which sexual divisions were derived. Darwin's theory of sexual selection, which he elaborated in the *Descent of Man* (1871), was founded on the assumption that sexual divergence was an integral part of the evolutionary process.[23] In common with many other contemporary bio-medical writers, Darwin held that the progenitor of the vertebrate kingdom was androgynous; he then argued that sex differences had begun to emerge in response to changing environmental conditions. A number of bodily structures and mental qualities not directly connected with the reproductive act had thus been developed, some of which gave the owner a reproductive advantage over other individuals of the same sex: this was the meaning Darwin attached to the term 'secondary sex characters'. These peculiarities had subsequently become established in the species through sexual selection, a mechanism independent of natural selection, which operated in two ways: either through the elimination of the weaker males in the contest for wives, or through the exertion of choice by the females over the courting males. Thus the males had acquired, for example, plumage, musical organs, strength and pugnacity – attributes which subserved either courtship or the struggle among the males for the possession of the females. The success of the better-endowed males in the contest for wives had gradually caused the male to diverge from the female; it was in this way, Darwin argued, that he had become her superior in terms of strength, pugnacity and mental powers.

Darwin's theory of sexual selection provoked great controversy within the scientific community. Some biologists insisted that sexual selection was a particular form of natural selection rather than a separate agency; others were shocked at the suggestion that birds and mammals possessed highly developed aesthetic tastes.[24] Criticism of Darwin's theory also came from those who held sexuality to be biologically determined. In the *Evolution of Sex*, for example, Geddes and Thomson argued that sex characters were not instruments to further reproductive success, but an expression of the 'fundamental physiological bias characteristic of either sex'. However, they did agree with Darwin's view that sexual dimorphism was a sign of 'organic progress' from a primitive hermaphroditic state: hermaphroditism in individuals should be interpreted either as a persistence of this condition, or as a reversion to it.[25]

THE INFLUENCE OF THE ENVIRONMENT

Although in the *Descent of Man* Darwin appeared to sanction the separation of men and women into distinct spheres of aptitude and ability, his insistence that sex differences were due to the action of the environment rather than of the reproductive organs had radical implications for theories of sexual division. If, as Darwin maintained, sexual dimorphism was the result of an evolutionary process which depended on the will, choice and rivalry of individuals, it might be possible consciously to alter the environmental cues through which sexual selection operated, so as to create a hermaphroditic race of feminine males and masculine females. This point was not lost on the British psychiatrist

Harry Campbell, a Darwinian and a supporter of the theory of sexual selection. As he observed in 1891, sexual selection showed one important fact – that some of the most prominent secondary sexual characters were 'not absolutely and inevitably necessary: matters might have been otherwise.' Given that woman was 'not what she is, and man what he is, simply because the one has ovaries and a uterus, and the other testicles', it was not inconceivable that

> *all* the secondary sexual characters in man and woman might be transposed – that the strength, courage, and fire of the man might be transferred to the woman; the weakness and timidity of the woman, to the man. For my part I have little doubt that such a transposition might be brought about in respect of most of the nervous differences by a process of artificial sexual selection carried on through many generations. In this connection it should be remembered that the secondary sexual characters are highly variable; whence arise abundant opportunities for the operation of selection.[26]

In this passage, Campbell acknowledged that neither femininity nor masculinity were inborn features fixed in 'nature': sexuality was the product of people's life history, and as such it could be consciously modified.

Interest in the environmental determinants of sexuality represents a prominent theme in the nineteenth-century debate on sex differences. Throughout the century it was widely assumed that sex characters, like any other physiological property of individuals, constantly interacted with a number of different external factors, such as geographical conditions, custom and mode of government.[27] Physiology and life-style affected each other: through habit and custom, the organism was modified by variables which were outside the human body. Thus close attention was paid to the influence of climate, diet and occupation on physiological processes like puberty and the menopause. For example, it was thought that a luxurious life-style induced puberty at an early age, whereas manual labour delayed it. Heat was said to promote menstruation, cold to check it, thus explaining why some women only menstruated in the summer.[28] When, in the mid-Victorian period, the pressure to open higher education to women began to mount, there was widespread anxiety within the medical profession that intellectual work would masculinize women. Many doctors anticipated that the educated woman would be muscular and angular – the loss of the mammary function and of the breasts themselves being regarded as the most likely consequences of academic study.[29]

Moving from the level of the individual to that of the species, the notion of civilization came to the fore in the work of a number of scientists as the means of explaining the evolution of certain sex characters in the animal kingdom. As the progress of society brought improvements to the human condition, changes occurred in people's reproductive biology, which marked off civilized man and woman from lower animals and primitive people. For example, the French zoologist Félix-Archimède Pouchet (1800–72), one of the chief proponents of the 'ovular theory' of menstruation, thought that the periodicity of menstruation in the human female had been induced by the easier living conditions which obtained in the state of civilization. Starting from a supposed analogy between menstruation in woman and the 'heat' or 'oestrus' in the lower animals, Pouchet argued that menstruation marked the spontaneous bursting of the ovarian follicle: menstruation thus coincided not only with the fertile period, but also with the peak of sexual desire in woman.[30] However, in the human female the frequency of the oestrus (and consequently her chances of being fecundated) had *increased*, thanks to the influence of civilization: living conditions in civilized societies were more favourable to the maintenance of offspring, Pouchet argued, hence civilization must have improved women's biological capacity to bear children. The same argument was advanced by Alfred Wiltshire in his 'Lectures on the comparative physiology of menstruation', published in 1883.[31] Writing in the light of Darwin's *Descent of Man* and *Variation of Plants under Domestication* (1868), in which Darwin examined the variations induced in animals and plants under changed environmental conditions, Wiltshire argued that the menstrual function was due to an increase in the frequency of the oestrus, and, like Pouchet, he attributed this change to the action of civilization.

Concern with the environmental aspects of sexuality often took the form of anthropological investigations into various aspects of sex and reproduction. These studies had meanings that went far beyond their explicit content: in the second half of the nineteenth century, the equation of the criminal, pauper and work-shy at home with the savage abroad formed a theme common both to anthropological speculation and to social investigations of life in the urban slums.[32] A typical example of this cross-cultural approach to sexuality is provided by the Manchester physician John Roberton (1797–1876), who interestingly carried out anthropological investigations into the age of menarche among Lancashire factory women as well as writing about the period of puberty in Eskimo women. The latter study focused on the effects of education and religious teaching on the inhabitants of Labrador.[33] Roberton started with a description of the geography of Labrador and of its inhabitants, whom he regarded as thievish, bloodthirsty and deeply degraded. He then noted that parturition had become more difficult since the arrival of European missionaries in the area. This, he argued, was due to the moral changes brought about by education and religious teaching, which had elevated Eskimo women to the rank of thinking beings while rendering their consitution more 'irritable' and susceptible to disorder.

However, it was not always easy to determine where the influence of culture ended and that of nature began. The obstetrician John Braxton Hicks (1823–97) clearly illustrates this difficulty in the series of lectures on sex differences he delivered in 1877. In his first lecture Hicks wondered whether sex differences were caused by changes in the primary sex organs at puberty, or whether both primary and secondary characters were 'the common result of a primary force extending to the whole body' during the embryonic phase.[34] In the second lecture, however, the balance shifted from nature to nurture. He noted that in terms of 'mental sensitiveness' men varied much amongst themselves, 'nearly, if not quite, as much as men differ from women', and he attributed men's greater control over their nervous system to 'their mode of bringing up, their more invigorating pursuits, their rougher contact with the world, . . . besides the differences of a similar kind derived by descent.' Exposure to the same social conditions eroded the differences between the sexes: 'Sooner or later,' Hicks argued, 'ill health, overwork, watching, anxieties, long-continued pain, failure in his pursuits, and many other things, singly or in combination, will bring man into a state so similar to that of woman under the same circumstances, that it must be acknowledged that it is only in degree that the sexes differ.'[35]

Hicks's uncertainties were echoed by other late-Victorian bio-medical writers. Commenting on the greater longevity and lower mortality rates of women, the surgeon W. R. Williams considered whether they were due to nature or culture. The higher death rates of males during infancy, 'when the dress, food, and general treatment of both sexes are alike', appeared to show that 'some constitutional condition inherent to sex' was probably at work. However, occupation also seemed to play a role, for in those countries where women engaged in hard labour, the mean duration of female life was considerably shorter than in England.[36] A similar thesis was advanced by the gynaecologist James Jamieson. While claiming that women's lower mortality rates during childhood was evidence for some constitutional property unique to the female, he also drew attention to the part played by civilization: 'There is good reason for believing', he wrote, 'that it is only in civilised communities that the average duration of life is greater in women than in men.'[37]

Set within this context, Havelock Ellis's conclusive chapter in *Man and Woman*, his comprehensive study of sex differences published in 1894, loses much of its radical edge for the modern reader: 'We have examined Man and Woman, as precisely as may be, from various points of view . . . ', he wrote at the end of his extensive review of the literature:

> It is abundantly evident that we have not reached the end proposed at the outset. We have not succeeded in determining the radical and essential characters of men and women uninfluenced by external modifying conditions . . . We have to recognise that our present knowledge of men and women cannot tell us what they might be or what they ought to be, but what they actually are, under the conditions of civilisation. By showing us that under varying conditions men and women are, within certain limits, indefinitely modifiable, a precise knowledge of the actual facts of the life of men and women forbids us to dogmatise rigidly concerning the respective spheres of men and women.[38]

Nevertheless, the recognition that the environment had such profound consequences for the characters of sex did not, on the whole, prevent bio-medical writers from assigning greater weight to biological factors in the psycho-physiological make-up of the female. As James Jamieson argued,

> The average male and female member of the human race resemble each other more closely at the extremes of life than in the middle period. Mere differences of habit and mode of life, which are most distinct during the adolescent and early adult periods, may go some way to account for the greater diversity then; but probably, it is chiefly due to the influence of motherhood, actual or potential, on the physical and mental economy of women. The reproductive function undoubtedly has a larger place, for good or evil, in the life of woman than in that of man.[39]

Thus, sex differences were not qualitative, but quantitative: they derived from the relative proportion of mutually opposed attributes in the individual. Woman was more physical, instinctual and emotional than man because more powerfully dominated by the sexual functions. By contrast, man's emotions and instinctual functions, including the sexual ones, were more firmly controlled by the brain: thus head injuries were liable to cause impotence and 'wasting of the testicles'; cretinism and lunacy arrested the development of the testes and suppressed the 'venereal appetite'.[40]

This different weighting of mental and physical events in the physiology of man and woman was reflected in the development of two methods of anthropological classification which enjoyed great vogue during the nineteenth century – namely, craniometry and pelvimetry, the study of cranial and pelvic capacity respectively. The use of brain size and cranial capacity as classificatory criteria dated back to the late eighteenth century, when comparative anatomists had begun to develop a number of indices, based on various skull measurements, by which they sought to discriminate between higher animals with very similar nervous systems: the size of the brain signified the degree of complexity of the nervous system, and this in its turn positioned organisms hierarchically on a spectrum of increasing perfection. During the first half of the nineteenth century, the use of cranial measurements came to be complemented by that of the pelvic index, a parameter devised by combining together the different diameters of the pelvis. In fact, the capacity of the pelvis was not irrelevant to the question of skull measurements, since it correlated with the size of the foetal head and the development of the intellectual faculties. But while craniometry was believed to be applicable to the classification of men, the pelvic index was thought to be a more reliable taxonomic criterion in woman. According to the cranial index, the white European male was invariably found to occupy the highest place in the order of races; the European woman, for her part, was more highly developed than her 'primitive' sister, except that her superiority was indicated by the greater capacity of her pelvis.[41]

Commenting on the significance of the pelvic index in their *System of Obstetric Medicine and Surgery* (1884), the gynaecologists Robert and Fancourt Barnes argued that pelvic differences between men and women originated from the functions the pelvis had to perform in each sex. The male pelvis was built for strength and 'powerful exertion'; the female's was modified for the sexual functions. However, some of the features of the pelvis were shared by both sexes: they were due to the erect posture, which was unique to the human species, thus differentiating the human pelvis from that of the lower animals. Taking the pelvic index as a measure of the progressive rise in the scale of mammalia, 'man, the noblest ape', showed 'an advance upon the gorilla', but this progress was not as marked as that of the female. Given that the 'highest pelvic type' was found in the European woman, was it legitimate to infer that man was the inferior animal? 'Man, perhaps, would appeal to another index', suggested the two gynaecologists, '– the cranium. If woman excels by the pelvis, man excels by the head.'[42] No comparison of inferiority or superiority could thus be made between the sexes: each was perfect in its own kind and could do in one direction what the other could not.

A SCIENCE OF FEMININITY

The quest for the biological foundations of femininity reached its climax in the 1840s, when research into the physiology of reproduction drew increasing attention to the role played by the ovaries. It then came to be widely held that the ovaries, the 'grand organs of sexual activity'[43] in woman, were the essential difference from which all others flowed. This was not surprising, since the sexual instinct was, by definition, the 'essence and the *raison d'être* of woman's form, the expression of the cause of her existence as a woman.'[44] In 1844 the French physician Achille Chéreau (1817–85) proposed to change Van Helmont's (1577–1644) dictum 'Propter solum uterum mulier est id quod est' into 'Propter solum ovarium mulier est id quod est.'[45] The same idea was vividly conveyed by Robert and Fancourt Barnes through a political metaphor, in which the ovary was depicted as a tyrant forever bending the female body to its imperious laws: 'In the ordinary state', the Barneses wrote in 1884,

> the active or dominant organ of the sexual system is the ovary. The reign of this organ is expressed by menstruation, the part taken by the uterus being secondary, or in obedience to the impulse of the ovary. The ovary reigns supreme until conception takes place; then the uterus succeeds, and rules until the child leaves it. Then it is deposed, and yields its place to the breast. The breast rules until it is more or less supplanted by the ovary, which is ever struggling for supremacy, and cannot long be kept in subjection.[46]

While the uterus and breast symbolized woman's maternal role, the ovaries were woman's tie with a subterranean world of drives and automatic behavioural responses over which she had no control.

It must be noted that feminists and medical men alike endorsed this view of femininity. When ovariotomy, a dangerous operation for the removal of cystic ovaries, became established in the second half of the nineteenth century, there was strong opposition to the procedure from feminists like Elizabeth Blackwell and Frances Power Cobbe, because it was thought to 'unsex' women. For this strand of feminism, women's claims to influence the public sphere derived from their 'natural' functions as mothers and guardians of the home, as Lynda Birke discusses elsewhere in this volume. The central problem with ovariotomy was that it purportedly destroyed woman's essence, thus threatening deep-seated beliefs about woman's nature, her role and social responsibilities. It was for this reason that throughout her life, Blackwell remained one of the most outspoken critics of gynaecological surgery.[47]

The conflation of the biological, social and moral aspects of women's sexual nature found its expression in the aesthetic appreciation of the female body: 'The relations of woman', claimed an anonymous writer on 'Woman in Her Psychological Relations' in 1851, 'are twofold; material and spiritual — corporeal and moral. By her corporeal nature she is the type and model of BEAUTY; by her spiritual, of GRACE; by her moral, of LOVE;' it was 'during the period of activity of the reproductive organs, peculiar to her physical construction, that the frame of woman is most pleasing and most beautiful.'[48] Of all the female characteristics, the beauty of the pelvis was underlined:

> It is in that portion of the body in immediate connexion with those parts peculiar to her organization, that the greatest beauty of form is found in woman, as though they were the *fons et origo* of corporeal as well as mental loveliness . . . The contours of the back are of the most admirable purity; the region of the kidneys is elongated, the scapulae scarcely visible; the loins grandly curved forwards, the haunches prominent and rounded; in short, the posterior surface of the torso in woman is unquestionably the *chef d'oeuvre* of nature.[49]

The 'encasing of the procreative organs and centre of procreative activity' embodied a 'divine Idea' of beauty and perfection equalled only by the bust, with its 'voluptuous contours and graceful inflexions': while the one was 'the manifestation of the instinct', the other expressed the sentiment of love which bound the mother to her offspring. Softness and roundness were taken to define the idea of beauty in woman, since these qualities differentiated living beings from inert matter, and symbolized woman's life-giving capacities. By contrast, sharpness and angularity, proper to inanimate objects, were antithetical to the idea of beauty. Thus menopausal and ovariotomized women, whose sexual functions were in abeyance, lost the softness and roundness of the female form, becoming the least beautiful and also the least moral women of all:

> With the shrinking of the ovaria . . . there is a corresponding change in the outer form. . . . The form becomes angular, the body lean, the skin wrinkled. The hair changes in colour and loses its luxuriancy; the skin is less transparent and soft, and the chin and upper lip become downy. . . . With this change in the person there is an analogous change in the mind, temper and feelings. The woman approximates in fact to a man, or in one word, she is a *virago*. . . . This unwomanly condition undoubtedly renders her repulsive to man, while her envious, overbearing temper, renders her offensive to her own sex.[50]

During the second half of the nineteenth century, the view that woman's body was entirely finalized for sex and reproduction provided the impetus for the definition of a new medical specialism devoted to the study of woman and her diseases. The science of gynaecology was far more than the investigation and treatment of the ailments which affected women's reproductive organs: it was a comprehensive inquiry into the physiological, mental, social and moral peculiarities that were deemed to result from woman's biological role. As the gynaecologist Lawson Tait wrote when he set out to account for the rise of his profession, 'the great function of a woman's life has for years made her the subject of specialists, male and female, the obstetricians. The subsidiary relations of her special organs and the special acquirements of her physique, based upon these, have necessitated the establishment of another class of specialist, the gynaecologist.'[51]

Tait was echoed by Barnes, who thus defined the scope of gynaecology in Quain's *Dictionary of Medicine* (1882):

> The word 'Gynaecology' . . . embraces far more than is expressed in the term 'diseases of women'. In its full etymological meaning it is comprehensive beyond the strict domain of medicine. . . . Without accepting the doctrine of Michelet, that the life of woman is a history of disease, it is undeniable that to appreciate justly the pathology of woman we must observe her in all her social relations, study minutely her moral and

intellectual characteristics – that we must, in short, never for a moment lose sight of those physical attributes which indelibly stamp her as a woman, which direct, control, and limit the exercise of her faculties. This collateral study is of infinitely more importance in the pathological history of woman than it is in that of man.[52]

Barnes did not agree with the view of femininity propounded by the French historian Michelet, for whom menstruation itself was an illness; however, he stressed that knowledge of 'natural woman' was the necessary foundation of gynaecological pathology and therapy. Woman's difference from man derived from the role 'nature' had assigned to her, hence feminine pathology could only be understood in the light of woman's total anatomy, physiology and social experience. While man's confrontation with the external environment explained his pathology, female disease was intelligible in terms of woman's role within marriage and the family: 'The integration of man with surrounding nature', wrote Barnes in 1880,

is more essential to the understanding of his physiology and pathology than in the case of woman. To the woman the integration with man and with offspring assumes predominant importance. Man's ambition and daily work is the contest with the surrounding world, physical and moral; the ambition of the work of woman, more restricted, is the study and conquest of man. This relation governs her being, and is the secret of a great part of her physiology and pathology.[53]

Starting from ideological assumptions about the dominance of the sexual functions in women, nineteenth-century gynaecologists proceeded to analyse the whole of the female organization by focusing on the mediating links between women's reproductive organs and their minds – be it the blood, a vestige of ancient humoral theories of disease, or the complex notion of reflex action, which significantly emphasized women's inability to exercise conscious control over their minds and bodies. This task involved an evaluation of the balance between instinct and reason, of the senses and the moral faculties, of the relationship between organization and environment – the very themes round which the natural history of Man, one of the chief components of the science of anthropology, was organized.

The attempt to construct a science that could explain woman's nature in its various physiological, moral and social aspects culminated in 1885 with the publication of Hermann H. Ploss's *Das Weib*, a book which quickly established its reputation as the standard work on woman and was subsequently updated and republished a number of times.[54] Ploss, a practising gynaecologist who founded the first midwifery clinic in Leipzig, envisaged his study as a 'natural history of woman' from puberty to childbirth: he intended it to illustrate the 'characteristic life and personality of woman' as revealed by anatomy, physiology, mythology and social custom. In order to achieve this aim, Ploss availed himself of the methods of gynaecology, anthropology and ethnography. *Das Weib* thus opened with an analysis of the physical and psychological characters peculiar to the 'sex' and continued with a review of the myths, legends and rituals which affected women's lives throughout the world. What gave this enterprise its meaning was the existence of certain assumptions about the nature of femininity: a vast array of races could be lumped together on the basis of 'sex', in the belief that every aspect of woman's physiology and social life demonstrated her specialization for the sexual functions.

This medico-anthropological discourse on gender contained an important silence, in that no science of masculinity analogous to gynaecology emerged during the Victorian period. Although men were liable to suffer from disorders of their sexual apparatus — such as inflammations of the prostate, impotence and hydrocele of the testes — in contrast to women they were not seen to be defined by their sexual pathology and physiology. Thus, although a specialism devoted to the treatment of men's genito-urinary problems did develop during the late Victorian period, it is perhaps no coincidence that attempts to redefine this study as the 'science of andrology' have been unsuccessful. In 1891 a group of urologists attending the Congress of American Physicians took the lead by forming themselves into the Section of Andrology, on analogy with gynaecology. This decision, which had been partly instigated by a desire to change the 'quack' image of the specialism, was not a happy one. Unfortunately for the urologists, the neologism was greeted with scorn and ridicule, and it subsequently fell into disuse.[55] Just over thirty years later, the English urologist Kenneth Walker proposed to separate the study of the diseases of the male organs of generation from those of the urinary tract; the new specialism, which was to be a definite branch of medicine comparable to gynaecology, was to be named 'andrology'.[56] As had been the case in the United States, the term 'andrology' did not catch on, and it was not until the late 1970s that this word was revived to designate a special branch of endocrinology — namely, the study of the hormonal determinants of masculinity.

CONCLUSION

Scientific notions of gender during the Victorian period reveal profound ambiguities about the nature of masculinity and femininity. Through ideas about the latent bisexuality of humankind, bio-medical writers expressed their belief in the continuity of male and female and their faith in the universality of human nature. Sex differences were not of kind, but of degree, and it was precisely the way in which one sex could shade into the other which posed the greatest problem for the Victorians.

In interpreting such ideas, it is tempting to refer to the theory of human nature which underpinned nineteenth-century liberalism. Laqueur has noted that the concept of gender difference is fundamentally at odds with the conception of personhood affirmed by liberal theory:

> liberal theory begins with a neuter body, sexed but without gender. . . . The body is regarded simply as the bearer of the rational subject, which itself constitutes the person. The problem for this theory then is how to derive the real world of male dominion of women, of sexual passion and jealousy, of the sexual division of labor and cultural practices generally from an original state of genderless bodies.

For Laqueur, the way out of this problem was through the construction of biological theories of sexual difference: 'The dilemma, at least for theorists interested in the subordination of women, is resolved by grounding the social and cultural differentiation of the sexes in a biology of incommensurability that liberal theory itself helped bring into being.'[57] My point is that it was possible to subscribe to idealized notions of personhood without even questioning the realities of gender inequality. The concept of latent hermaphroditism allowed bio-medical writers to do just that: within the deepest recesses of their bodies, men and women were neuter persons, undifferentiated parts of a society which in theory recognized the equality of all its members. One might mention in this connection the role played by androgyny in the political cosmology of a number of early nineteenth-century socialist and millenarian groups. For example, in the writings of John Goodwyn, one of the founders of the Communist Church, the history of humankind was characterized by periodical swings between the forces of 'man-power' and 'woman-power'; the point of 'equilibration' was reached in the birth of individuals who contained within themselves both masculine and feminine principles.

Eventually the whole earth was to be populated by this androgynous race, created in socialist communities where 'unsexual Chartism' and female emancipation would be assured.[58]

During the twentieth century, theories of hormonal functioning have come to govern our understanding of sexual differentiation, but the central issues examined in this chapter remain unresolved: as Victor Medvei writes in his history of endocrinology, the seeming 'ambivalence of sexuality' continues to puzzle contemporary endocrinologists. Medvei lists several examples. In 1934, E. A. Haeussler found that the testes of stallions contained five-hundred times more oestrone than the ovaries of sexually mature mares. Oestrone is present in men's urine; vice versa, the amount of androgenic substances found in women's urine is similar to that found in men's. Quoting the eminent endocrinologist Sir Alan S. Parkes, Medvei thus comments: 'The present wonder, therefore, is not that intersexual conditions occur, but that the balance of endocrine factors usually comes down on one side or the other to produce a recognisable male or female.'[59] But perhaps at this point even the endocrinologist might begin to wonder what it is in our culture that makes a human being a 'recognisable' male or female.

ACKNOWLEDGEMENTS

This chapter is based on material which appears in O. Moscucci, *The Science of Woman: Gynaecology and Gender in England, 1800–1929* (Cambridge University Press, Cambridge, 1990). I wish to thank Marina Benjamin for her help in preparing this essay.

NOTES

1 E. Clarke, *Sex in Education; Or, a Fair Chance for the Girls* (J. R. Osgood, Boston, 1873), p. 12.

2 See E. Fee, 'Science and the "Woman Question": a study of English scientific periodicals' (Ph.D. thesis, Univesity of Princeton, 1978); E. K. Helsinger, R. L. Sheets and W. Veeder (eds), *The Woman Question: Social Issues* (3 vols, Manchester University Press, Manchester, 1983). For the feminist contribution to the debate, see R. Rosenberg, *Beyond Separate Spheres* (Yale University Press, New Haven and London, 1982).

3 J. Jamieson, 'Sex, in health and disease', *Australian Medical Journal*, n.s. 9 (1887), p. 146.

4 T. Laqueur, 'Orgasm, generation, and the politics of reproductive biology', *Representations*, 14 (1986), pp. 1–41.

5 L. Jordanova (ed.), *Languages of Nature: Critical Essays on Science and Literature* (Free Association Books, London, 1986), p. 35.

6 On the 'science of Man', see S. Moravia, *Filosofia e scienze umane nell'età dei lumi* (Sansoni Editori Nuoba S.p.A., Florence, 1982); M. S. Staum, *Cabanis: Enlightenment and Medical Philosophy in the French Revolution* (Princeton University Press, Princeton, 1980).

7 P. J. G. Cabanis, *Oeuvres Philosophiques* (2 vols, Presses Universitaires de France, Paris, 1956), 1, p. 275.

8 J.-J. Rousseau, *Émile: Or, on Education*, book 5 (Basic Books, New York, 1979, pp. 357–8.

9 The emergence of this discourse in France is examined by Y. Knibiehler, 'Les médecins et la "nature féminine" au temps du code civil', *Annales: E.S.C.*, 31 (1976), pp. 824–45.

10 P. Roussel, *Système physique et moral de la femme*, 2nd edn (1777), (Crapart, Caille and Ravier, Paris, 1803); W. Alexander, *The History of Women, from the Earliest Antiquity, to the Present Time; Giving Some Account of Almost Every Interesting Particular Concerning That Sex, among All Nations, Ancient and Modern* (T. Strahan and T. Cadell, London, 1779).

11 G. J. Romanes, 'Mental differences between men and women' in Romanes, *Essays*, ed. C. Lloyd Morgan (Longmans, London, 1897), p. 113.

12 R. Barnes, 'On vicarious menstruation', *British Gynaecological Journal*, 2 (1886–7), pp. 151–83.

13 V. O. King, 'Case of menstruation in the male', *Canada Medical Journal*, 2 (1867), p. 472.

14 A. Wiltshire, 'Clinical lectures on vicarious menstruation, or menses devii', *Lancet*, 2 (1885), pp. 513–17.

15 T. Laycock, *A Treatise on the Nervous Diseases of Women* (Longman, Orme, Brown, Green and Longman, London, 1840), p. 218.

16 Wiltshire, 'Vicarious menstruation', p. 516.

17 Quoted in F. J. Sulloway, *Freud, Biologist of the Mind: Beyond the Psychoanalytic Legend* (Basic Books, New York, 1979), p. 140.

18 A table of homologies is provided in J. Y. Simpson, 'Hermaphroditism', in W. G. Simpson, *The Works of Sir J. Y. Simpson* (3 vols, A. and C. Black, Edinburgh, 1871), 2, pp. 509–10.

19 A. Farre, 'Uterus and its appendages' in R. B. Todd (ed.), *The Cyclopaedia of Anatomy and Physiology* (5 vols, Longman, Brown, Green, Longman and Roberts, London, 1835–59), 5, p. 623.

20 Simpson, 'Hermaphroditism', pp. 491–2.

21 Ibid., p. 490.

22 Ibid.

23 C. Darwin, *The Descent of Man and Selection in Relation to Sex* (1871), 2nd edn, (J. Murray, London, 1909).

24 C. E. Russett, *Sexual Science: The Victorian Construction of Womanhood* (Harvard University Press, Cambridge, Mass., 1989), p. 94.

25 P. Geddes and J. A. Thomson, *The Evolution of Sex* (Walter Scott, London, 1889), p. 67.

26 H. Campbell, *Differences in the Nervous Organisation of Man and Woman: Physiological and Pathological* (H. K. Lewis, London, 1891), pp. 46–7.

27 Concern with the environmental aspects of health and disease went back to the times of Hippocrates. A discussion of environmentalism in medicine during the late eighteenth century is provided by L. Jordanova, 'Earth science and environmental medicine: the synthesis of the Enlightenment' in L. Jordanova and R. Porter (eds), *Images of the Earth* (The British Society for the History of Science, Chalfont St Giles, 1979), pp. 119–46.

28 See, for example, R. Barnes and R. F. Barnes, *A System of Obstetric Medicine and Surgery* (2 vols, Smith, Elder, London, 1884), 1, p. 42.

29 See, for example, Clarke, *Sex in Education*.

30 F. A. Pouchet, *Théorie positive de l'ovulation spontanée et de la fécondation des mammifères et de l'espèce humaine* (J.-B. Baillière, Paris, 1847), pp. 233, 244.

31 A. Wiltshire, 'Lectures on the comparative physiology of menstruation', *British Medical Journal*, 1 (1883), pp. 395–8.

32 See, for example, Henry Mayhew's treatment of London's 'social residuum' in *London Labour and the London Poor: A Cyclopaedia of the Conditions and Earnings of Those That Will Work, Those That Cannot Work, and Those That Will Not Work* (4 vols, Griffin, Bohn, London, 1861).

33 J. Roberton, 'On the period of puberty in Esquimaux women', *Edinburgh Medical and Surgical Journal*, 63 (1845), pp. 57–65.

34 J. B. Hicks, 'On the differences between the sexes in regard to the aspect and treatment of disease', *British Medical Journal*, 1 (1877), pp. 318–20, 347–9, 377–9, 413–15, 447–9, 475–6, p. 319.

35 Ibid., p. 414.

36 W. R. Williams, *The Influence of Sex in Diseases* (J. and A. Churchill, London, 1885), p. 3.

37 Jamieson, 'Sex', p. 152.

38 H. Ellis, *Man and Woman: A Study of Human Secondary Sexual Characters* (1894), 4th edn (Walter Scott, London, 1904), p. 223.

39 Jamieson, 'Sex', p. 146.

40 T. B. Curling, 'Testicles', in Todd, *The Cyclopaedia of Anatomy and Physiology*, 4, part ii, pp. 985, 992, 994.

41 On pelvimetry, see G. Vrolik, *Considerations sur la diversité des bassins de*

differentes races humaines (J. Van der Hey and Sons, Amsterdam, 1826); J. G. Garson, 'Pelvimetry', *Journal of Anatomy and Physiology*, 16, (1881–2), pp. ·106–34. The gendering of anatomy from the late eighteenth century onwards is discussed by L. Schiebinger, 'Skeletons in the closet: the first illustrations of the female skeleton in eighteenth-century anatomy', *Representations*, 14 (1986), pp. 42–82.

42 Barnes and Barnes, *System of Obstetric Medicine*, 1, pp. 170–3.

43 C. West, *Lectures on the Diseases of Women*, 3rd edn (J. Churchill and Sons, London, 1864), p. 5. The first proponents of the 'ovular theory' of menstruation were J. Power, R. Lee, T. L. W. Bischoff, A. Raciborski and F.-A. Pouchet. For a discussion, see H. H. Simmer, 'Pflüger's nerve reflex theory of menstruation: the product of analogy, teleology and neurophysiology', *Clio Medica*, 12 (1977), pp. 57–90; Laqueur, 'Orgasm', esp. pp. 24–32.

44 W. Balls-Headley, *The Evolution of the Diseases of Women* (Smith, Elder, London, 1894), p. 1.

45 A. Chéreau, *Memoires pour servir à l'étude des maladies des ovaires* (Jortin, Masson and Cie, Paris, 1844), p. 91.

46 Barnes and Barnes, *System of Obstetric Medicine*, 1, pp. 202–3.

47 O. Moscucci, *The Science of Woman: Gynaecology and Gender in England, 1800–1929* (Cambridge University Press, Cambridge, 1990), esp. ch. 5. On the connections between feminism, antivivisection and the campaign against gynaecological surgery, see also M. A. Elston, 'Women and antivivisection in Victorian England' in N. Rupke (ed.), *Vivisection in Historical Perspective* (Croom Helm, London, 1987), pp. 259–94.

48 'Woman in her psychological relations', *Journal of Psychological Medicine and Mental Pathology*, 4 (1852), pp. 18–50, esp. pp. 18–19.

49 Ibid., p. 20.

50 Ibid., p. 35.

51 R. L. Tait, *Diseases of Women and Abdominal Surgery*, vol. 1 (no further vols published), (Richardson, Birmingham, 1889), p. 3.

52 R. Barnes, 'Women, diseases of' in R. Quain (ed.), *A Dictionary of Medicine: Including General Pathology, General Therapeutics, Hygiene, and the Diseases Peculiar to Women and Children* (2 vols, Longman, Green, London, 1882), 2, p. 1789.

53 R. Barnes, 'Lectures on the diseases of women', *Lancet*, 1 (1880), pp. 155–7, at p. 156.

54 H. H. Ploss, M. Bartels and P. Bartels, *Woman: An Historical, Gynaecological and Anthropological Compendium*, ed. E. J. Dingwall (3 vols, Heinemann, London, 1935).

55 'Andrology as a specialty', *Journal of the Americal Medical Association*, 17 (1891), p. 631.

56 K. M. Walker, *Diseases of the Male Organs of Generation* (H. Frowde, and Hodder and Stoughton, London, 1923).

57 Laqueur, 'Orgasm', p. 19.

58 B. Taylor, *Eve and the New Jerusalem: Socialism and Feminism in the Nineteenth Century* (Virago, London, 1983).

59 V. C. Medvei, *A History of Endocrinology* (MTP, Lancaster/Boston/The Hague, 1982), p. 406.

7

Hysteria Male/Hysteria Female: Reflections on Comparative Gender Construction in Nineteenth-Century France and Britain

Mark S. Micale

Over the past ten years or so, professional historical studies have witnessed a burst of publications on the theme of gender. It is scarcely possible today to visit an academic bookstore or open a copy of a review journal without encountering a new biography or monograph, a collection of essays or conference announcement devoted to the subject. Recently, this scholarly interest has extended to the subject of gender and the history of science and medicine. Within the humanities, it is fair to say that the intersection of gender, science, and medicine is among the most promising 'research sites' today. However, to date, the overwhelming majority of the scholarship has focused narrowly, and quite separately, on the historical experience of women. This is particularly true in the field of of the history of psychiatry, where the issue of gender has emerged as practically synonymous with the treatment of women in past psychiatric theory and practice.[1] Thus far, the 'new men's studies' have been limited largely to topics in general social and cultural history.[2] Furthermore, almost no attention has been paid to one of the most significant aspects of the field, namely, the *comparative* study of the two sexes in historically specific settings. It was in 1985 that Joan Scott, building on remarks of Natalie Davis, noted at the annual meeting of the American Historical Association that gender studies 'introduces a relational notion into our analytic vocabularies. According to this new view, women and men are defined in terms of one another, and no understanding of either can be achieved by entirely separate study.'[3] The challenge perceived by Davis and Scott, that is, the formulation of a truly *interactionist* model of gender and history, represents, it seems to me, one of the major conceptual gains in the movement from the women's studies of the 1960s and 1970s to the gender studies of the present. Several years after these comments, however, the subject remains to be explored by historians of the medical, and particularly the mental sciences.

The study of the history of hysteria nicely illustrates the problems and potentialities of the comparative historical study of gender. In the case of hysteria, a focus on the female sex at first appears inevitable. If there is a psychodiagnostic category that we associate overwhelmingly with women, it is hysteria. Etymologically, the term traces to the Greek *hystera*, or uterus. The Hippocratic doctrine of the wandering womb is well known. Many current-day historians and psychiatrists believe that women persecuted for witchcraft during the early modern period displayed behaviours that were actually hysterical in nature. And the great majority of medical writings on the neurosis over the past three centuries have dealt with the disorder as an affliction of adult and adolescent women.[4] On first consideration, the nineteenth century appears no exception to the rule. In the popular historical mind today, the image of hysteria during its famous *fin de siècle* phase is of a phenomenon thoroughly female, and by now we have a significant scholarly literature on the association of psychiatry, nervous disease, and the nineteenth-century bourgeois woman.[5] By and large, our knowledge of this subject has derived from a number of memorable and widely-read documents, such as Charlotte Perkins Gilman's *The Yellow Wallpaper*, the letters and diaries of Alice James, the *Iconographie photographique de la Salpêtrière* of Charcot, and Freud's and Breuer's book of 1895, *Studies on Hysteria*, in which all of the patients are women.

However, if we move beyond canonical literary and medical texts to a consideration of all available medical-historical sources, we find a substantial body of evidence to suggest that men and children too were implicated in the classically Victorian forms of psychopathology. As Howard Feinstein has observed, the debilitating neurasthenic invalidism that occurred so often in middle- and upper-middle class women in the nineteenth century struck as well figures such as Charles Darwin, William James, Louis Agassiz, and William Dean Howells.[6] Similarly, the British medical literature from the 1880s on so-called 'railway spine', and from the following decades on 'traumatic neurasthenia,' dealt with men, women, and children alike. In the United States, Silas Weir Mitchell derived his famous rest-cure from therapeutic techniques developed during the American Civil War for battle fatigue among soldiers.[7] And Francis Gosling's recent quantitative study of the sizeable American body of writing on neurasthenia demonstrates that George Beard's fashionable diagnosis was applied with rough equivalency to males and females.[8] Similarly the on-going research of Edward Shorter on private nervous clinics in Austria at the turn of the century is revealing a clientele for some clinical facilities of more men than women.[9] A majority of Freud's major clinical writings – the cases of Little Hans, the Rat Man, Schreber, and the Wolf Man – involved men, while the most famous analysand of all, Freud himself, was, of course, male. And finally, as numerous scholars have shown of late, a quantity of medical writing on shell-shock and the war neuroses emerged during the Russo-Japanese War of 1905 and the First World War.[10] No less than nineteenth-century medical commentary on women, these literatures concerning sick and suffering men reveal normative gender representations, encoded ideals of normal and abnormal masculinity that repay investigation by the social, cultural, and medical historian.

Perhaps most significant in this regard is the French-language literature on hysteria in the male. A review of primary medical-historical sources reveals at once that the final two decades of the nineteenth century were marked by a spirited controversy over the question of hysteria in adult and adolescent men. The prime mover in this debate was the Parisian neurologist Jean-Martin Charcot. During a 12-year period that corresponds roughly with the 1880s, Charcot published the case histories of more than 60 male 'hysterics' and treated countless others in his daily hospital practice. In these writings, Charcot elaborated a set of ideas about the disease in males, including a theory of etiology, a model of symptomatology, and a therapeutic programme. Through his work on hysterical disorders in males, Charcot accomplished some of his most substantial scientific work, concerning subjects such as the ideogenesis of functional nervous disorders and the differential diagnosis of hysterical and organic syndromes. Moreover, as the leading figure in the 'School of the Salpêtrière' Charcot's writings inspired a large periodical and monographic literature on hysteria in the male, which appeared during the 1880s and 1890s in all of the major European languages.[11] As the editor of a recent collection of Charcot's case histories observes, 'The volume of these writings [on male hysteria], the importance that Charcot attached to them, [and] their significance for the interested public at the time, was not less than that which attended the writings on feminine hysteria several years before.'[12] The appearance of a large body of medical commentary on male hysteria alongside Charcot's well-known work on hysteria in the female offers, in the manner suggested by Davis and Scott, an invaluable opportunity to explore the process of comparative gender construction a century ago.

CHARCOT AND THE THEORIES OF HYSTERIA IN THE MALE AND THE FEMALE: THEORETICAL SIMILARITIES

Charcot conducted his initial investigation of hysteria at the Salpêtrière Hospital on the south-eastern edge of Paris during the period 1872–8. Since its inception in the mid-seventeenth century, the Salpêtrière had been reserved for female patients, while their male counterparts were shipped further outside the city to Bicêtre. During this six-year period, Charcot 'constructed' his basic nosographical model of the disease, a model, it is important to realize, based entirely on the observation of female patients. With a longstanding and instinctive distrust of medical philosophizing, Charcot never produced a general theoretical study of hysteria, in either its masculine or feminine versions. Rather, his writings took the form of compilations of case histories that were intended to communicate through lucid clinical illustration and to convince the reader through the sheer accumulation of empircal data. In formulating his proto-theory of hysteria during the 1870s, Charcot produced seven major clinical histories concerning female hysterical patients, which appeared in the first volume of his *Leçons sur les maladies du système nerveux*, as well as numerous other cases in medical periodicals.[13]

In 1882, specifically at Charcot's instigation, there opened on the grounds of the Salpêtrière a new *Service des hommes*. This consisted of a special 20-bed ward in the General Infirmary of the hospital for males suffering from transient nervous or neurological disorders.[14] The first writings in Charcot's *œuvre* on hysteria in males date from this period.[15] Charcot's acceptance of male patients at this age-old sanctum of female pathology neatly parallels his bringing of the ancient 'feminine' diagnosis of hysteria to the male half of the population. From the early 1880s until his death in 1893, Charcot wrote steadily, and simultaneously, on hysterical disorders in both its masculine and feminine forms. From my tabulations, his overall bibliography includes 61 published cases of males of all ages and 73 females who received the primary diagnostic labels of 'hystérie', 'grande hystérie,' 'hystéro-épilepsie,' or 'hystéro-neurasthénie.'[16] The Bibliothèque Charcot in Paris houses today the clinical notes, many in Charcot's own hand, for another 30 cases involving men, and for over 220 cases of female hysteria.

In general structure, the disease picture that emerges from Charcot's collected case-historical writings is similar for male and female patients. Charcot conceptualized hysteria in both men and women as a hereditary degenerative disorder, caused by a combination of underlying constitutional factors and short-term triggering mechanisms, manifesting a wide range of physical (especially neurological) symptoms, with an unpredictable pattern of evolution, and a largely unfavourable prognosis. In members of both sexes, the disorder was also often characterized by an epileptiform attack. Charcot believed that the attack could be started and stopped through the application of pressure or stimulation to various 'hysterogenic zones' on the body of the patient. Many times in the course of his writings, Charcot insisted on the essential congruence of hysterical disorders in the two sexes.[17] This is especially true in his early cases of male hysteria appearing from 1882–5 in which he was primarily concerned simply to establish the widespread reality of hysterical disorders in males. A favourite argumentative device to this end was the juxtaposition in lecture of male and female patients, with Charcot reviewing for the audience the parallel symptomatologies. Especially noteworthy in this regard are the seventh and eighth lectures in the third volume of the *Leçons sur les maladies du système nerveux* in which he presented two male patients beside Blanche Wittman, celebrated the 'Queen of the Hysterics.'[18]

Taken in its entirety, however, Charcot's work represents much more than the simple and direct application of an age-old diagnostic category to members of the opposite sex. Close analysis in fact reveals a complex series of resemblances and differences in the Charcotian models of male and female hysteria. Regarding similarities, Charcot's representation of hysterical sickness in men and women began with the deconstruction of traditional ideas and attitudes about the disorder, and it was on this point, in its critical attitude toward anterior medical doctrine, that his ideas on hysteria in the two sexes are most alike. From the start, Charcot rejected strenuously all genital etiologies of hysteria. From Graeco-Roman medicine to psychoanalytic theory, hysteria was linked in one way or another to inadequate or excessive or impaired sexuality. By Charcot's time, the literal doctrine of the wandering womb had long been discredited; but gynaecological theories of the disorder continued to persist in various anatomically and physiologically updated versions throughout the nineteenth century.[19] Charcot believed that these old associations were erroneous and harmful, and he argued continuously against any possible sexual pathogenesis of hysteria. (As we will see shortly, however, what he excluded from the etiological realm he was willing to re-admit in the area of descriptive symptomatology.) In a majority of Charcot's clinical reports, then, sexual matters are not mentioned at all. In the few cases where they do figure, they invariably assume the auxiliary states of *agents provocateurs*, most often in the form of physical disorders or dysfunctions of the urogenital system. In his cases involving women, Charcot makes very little of menstruation, the factor that Griesinger and many other nineteenth-century physicians saw as the great irritant of women's nervous problems. And his writings rarely include reference to menopause, pregnancy, or postpartum conditions as causal factors. To similar effect, in only three of his over five-dozen cases involving hysterical males does he mention masturbation, a striking fact given the great medical preoccupation at the time with the perils of self-pollution.[20]

Charcot's scepticism of sexual theories of hysteria was accompanied by his dismissal of the nineteenth-century doctrine of 'innate dispositions.' Typically, Victorian psychological medicine located the primary cause of female nervous afflictions in the basic physical and biological make-up of women. In this view, women, in order to fulfil their natural domestic and maternal roles, were endowed with special sensitive and emotional natures, while men were equipped for the rough external world with a much sturdier nervous apparatus. As Elaine Showalter has observed, this line of thinking, which provided the medical underpinning for the doctrine of the separate spheres, was built on 'an ideology of absolute and natural difference between women and men.'[21] Within the intellectual history of hysteria, this ideology took the form of the very Victorian notion of a 'hysterical constitution.' Hysteria here was seen to represent less a disease entity than a kind of pathological intensification of female nature itself. 'As a general rule,' wrote Augustin Fabre, Professor of Clinical Medicine at Marseille, 'all women are hysterical and . . . every woman carries within her the seeds of hysteria. Hysteria, before being an illness, is a temperament, and what constitutes the temperament of a woman is rudimentary hysteria.'[22] If women are truly to be hated, berated, and controlled, it is most efficient to attribute their inferiority to innate causes. Accordingly, the most overtly misogynistic theories of hysteria in the *fin de siècle* – those of Möbius, Weininger, and certain German gynaecological surgeons – drew heavily on the concept of a female hysterical predisposition.

Now, despite the fashionability of the idea of nervous and mental constitutions, and its considerable serviceability in explaining the therapeutic impotence of doctors confronted with rampant nervous ailments among their female patients, Charcot took a stand against the concept of innate sex differences. In its place, he turned for a scheme of primary etiology to contemporary French degeneration theory. Following Morel, Moreau, and later Magnan, Charcot believed that hysterical patients possessed from birth a latent flaw or lesion of the nervous system. This *tare nerveuse*, he claimed, was inherited directly or indirectly from defective parental stock and stood ready at all times to be activated by appropriate environmental circumstances.[23] Degeneration theory, which reached a high point of influence in the French medical and biological sciences during the 1880s, was terrible in its own right. But nineteenth-century degenerationism offered a *neurological* rather than anatomical or physiological determinism and as such was much less exclusive than alternate theories in its gender associations. Earlier gynaecological and characterological theories had served equally to link hysteria, exclusively and negatively, with women. The *diathèse nerveuse*, however, occurred commonly in men, women, and children.[24]

Related to these beliefs were Charcot's arguments against what he considered the most enduring and damaging gender stereotypes associated with hysteria. Again, this concerns his attitude toward both male and female patients, although precisely what the stereotypes consisted of differed between the sexes. Later in the century writings on the hysterical temperament presented female patients as deceitful, immature, emotionally manipulative, and sexually-provacative creatures.[25] This was a prejudice that found its remote origin in the sixteenth- and seventeenth-century literature on witchcraft, and that received powerful restatement in France during the 1870s and 1880s in a medical literature on hysteria and nymphomania.[26] Charcot, however, believed that this image demeaned the suffering patient and, more significant for him, prevented the serious scientific study of an important medical condition. As a consequence, we find him repeatedly calling into question the alleged libidinous or lascivious character of the hysterical woman.[27]

With male hysterics, Charcot avoided above all the spectre of gender ambivalence. Anticipating astutely the areas of professional resistance to recognition of hysteria in men, Charcot argued against the ideas that the disorder was limited to boys at the age of puberty; that it existed solely in the leisured upper classes of society; that it appeared mainly in bachelors or priests; and that it occurred only in effeminate or homosexual men. Accordingly, Charcot is careful throughout his writings to establish that male hysteria – or 'virile hysteria' as he liked to call it – is authentically masculine in nature.[28] Coming from the neighbourhoods surrounding the Salpêtrière, nearly all of his patients were working-class men: masons, bakers, carpenters, gardeners, plumbers, locksmiths, railway employees, factory workers, and so on. In many cases, Charcot made a point of establishing the solid heterosexual credentials of his patients. 'Those adult men who are prey to the hysterical neurosis,' he insisted in 1884

> do not always present characteristics of effeminacy. Far from it. They are in the majority of cases robust men presenting all the attributes of the male sex, soldiers or artisans, married and the fathers of families, men, in other words, in whom one would be surprised, unless forewarned, to meet with an illness considered by most people as exclusive to women.[29]

Unlike Mitchell, George Beard, and later Freud, Charcot chose not to integrate clinical materials from his lucrative private practice into the theory of hysteria. Indeed in numerous passages, he seems to offer direct contrasts between the sturdy and reliable manliness of his working-class patients and 'les sujets plus délicat, plus impressionnables des classes lettrées.'[30] On this point, then, we find a complementarity of socio-economic character and gender identity in the patients selected by Charcot. Whether the picture that emerges in these writings of male gender identity is an accurate portrayal of male working-class masculinity, or a picture of what professional middle-class diagnosticians thought male working-class gender should look like, is an interesting matter for speculation. The significant point is that while Charcot was able to break with tradition in the full, he believed equal, recognition of hysterical disorders in men and women, he went out of his way, at times to the point of excluding from publication disconfirming clinical evidence, to preserve the traditional gender identities of hysterics of both sexes.[31]

THEORETICAL DIFFERENCES:

In many other ways, the Charcotian theories of hysteria for males and females differed significantly. A number of these differences Charcot acknowledged, while others he did not. Concerning causal theory, we have seen that Charcot posited a primary hereditarian origin of hysteria. However, in applying a general Morelian model of nervous disease to hysteria, he formulated two sub-concepts that were conspicuously unalike for males and females. There was first the comparative ratio of occurrence of the disorder. Following Pierre Briquet, Charcot believed that hysteria occurred between the sexes in a ratio of roughly 1:20.[32] This meant that only about 5 per cent of the total population of hysterical individuals was male. Second, according to Charcot's etiological reasoning, the maternal contribution to the transmission of hysteria was direct, the paternal indirect. In other words, a father with a degenerative disorder, such as alcoholism, epilepsy, or syphilis, could, by means of 'transformational heredity,' generate hysteria in a daughter or son; but the disorder could be conveyed directly, without change in form, only from the mother to a male or female child.[33] Behind every hysterical child, the suspicion lingered, there stood a hysterical mother. Both of these ideas, it seems to me, reflect the enduring belief that *au fond* hysteria remained a female affliction.

The greatest area of divergence in Charcot's thinking about male and female hysteria concerned the theory of secondary or 'exciting' causes. If the primary operation of heredity was uniform and constitutional, secondary causes were innumerable and circumstantial and therefore more open to the reflection of a range of subjective 'non-scientific' factors. Among the most significant of these factors was gender. In 38 out of Charcot's 73 published case histories of female hysteria, or just over one-half, a hereditarily tainted patient develops hysterical symptoms as the result of an overpowering emotional experience. In order of frequency, these experiences involved marital turmoil, unrequited love, death of a family member, religious ecstasy, and superstitious fear. The operative emotions in these events were anxiety, unhappiness, enthusiasm, fear, and grief. And in Charcot's narratives the events that elicit these sentiments transpired most often in domestic settings. By contrast, the largest percentage of Charcot's 61 cases of hysteria in men are set in motion through the destructive influence of a physical trauma, most often a trauma occurring in a public workplace: a blacksmith burns his hand and forearm with a hot iron; a clerk in an oil factory is nearly crushed to death by falling metal storage-barrels; a bricklayer falls two floors from his scaffolding; a ditch-digger is struck in the face with a shovel while unloading a wagon.[34] These experiences were often compounded by prior venereal infection or alcoholic excess. Charcot's male cases rarely occurred outside the context of a direct bodily threat, and the female cases rarely appeared in association with alcoholic or venereal factors. When hysteria in males was precipitated by a purely emotional force – and this occurred in only 6 out of 61 cases, or 10 per cent – it was elicited most often by such 'manly' emotions as rage, jealousy, and agitation. In other words, it appears that, knowingly or unknowingly, Charcot formulated for the two sexes an essentially separate set of secondary causal factors, factors that were consonant with prevailing notions of masculine and feminine natures. Simplifying somewhat, we can say that women in his writings fall ill due to their vulnerable emotional natures and an inability to control their feelings. In contrast, men get sick from working, drinking, fighting, and fornicating too much. However, in members of both sexes, it should be noted, hysteria typically resulted from an *excess* of prescribed gender behaviours.

We can detect a similar gender-specific pattern in other areas of Charcot's theorization. Consider his model of symptomatology. Charcot believed that hysteria was a *névrose*, but a neurosis in the strictly nineteenth-century neurological sense of the term. Accordingly, the defining characteristics of the illness were signs of central nervous system dysfunction, such as sensory anaesthesias and hyperaesthesias, abnormalities of vision, hearing, touch, and taste, and paralyses and contractures.[35] This is again true for patients of both sexes. In Charcot's female patients, these classic neurological somatizations are commonly accompanied by symptoms such as extravagant mood shifts, attacks of anxiety, fits of crying, and threats of suicide. However, Charcot's cases involving hysterical males are rarely complicated by such unseemly displays of emotionality. By and large, Charcot provided remarkably little information about the *mental* state of his patients. His case histories involving males consist in the main of narrations, dense and technical in nature, of the purely physical manifestations of the disorder. If he reported an extreme mental condition in a male patient, it was usually depression. In those instances where sexual or psychological behaviours were undeniable, he was likely to employ the mixed diagnostic label 'hystero-neurasthenia.'[36]

In Charcot's over five-dozen case histories, there are only two instances of an adult man crying.[37] 'The hysterical men of the working class who . . . fill the hospital wards of Paris today,' he observed in one of his *Leçons du mardi*, 'are almost always sombre, melancholic, depressed and discouraged people. . . . We should not expect to find in the male that morbid *con brio* frequent in reality in the female.'[38] In retrospect, it is likely that Charcot's male patients experienced the same range of emotions as their female counterparts; yet Charcot seems to have 'seen' and recorded only the most 'objective' externalized aspects of the disorder, while ignoring those symptoms we would today consider psychotic or psychoneurotic.

We get a still closer view of this pattern of theoretical behaviour by examining a single element of the symptomatological model, the hysterical attack. Charcot believed that men and women equally were subject to hysterical fits, and his writings offer several dramatic and detailed accounts of hysterical attacks undergone by patients of both sexes.[39] Charcot's *grande attaque hystérique* consisted of four clearly demarcated stages: an epileptoid period, marked by tonic and clonic muscular spasms; a stage of *grands mouvements*, in which the patient assumed striking and stylized postures, such as the arched pelvis position; a phase of *attitudes passionnelles*, characterized by the hallucinatory re-enactment of emotional scenes from the patient's past; and a lengthy and delirious period of withdrawal. From among these stages, the third phase, of *attitudes passionnelles*, was the most intense psychologically. In his descriptions of hysterical seizures in women, these four stages receive roughly equal attention. The provocative passional states of fear, pleasure, surprise, and religious and erotic ecstasy are documented extensively. But there is again a sharp and instructive contrast with Charcot's imaging of the male seizure. Charcot claimed that hysteria in the male took, if anything, a more convulsive form than in the female. A slightly larger percentage of male cases involved convulsive phenomena, and these behaviours tended to assume more violent and acrobatic forms than in females. However, the scope of emotional expression represented in the third passional stage of the attack is noticeably narrower in adult men, and its emotional content tends decidedly toward the darker, depressive end of the spectrum. Moreover, in several cases involving males, the period of *attitudes passionnelles* is shortened or truncated, and in a substantial number of cases — 25 cases, or 41 per cent, to be exact — this stage of the attack is absent altogether.[40] In place of the hyperemotionality of the third period, we read much more in the clinical reports concerning men about the first, epileptiform period of the attack. The pattern is again clear. The diagnosis has been drained of its affective content, and the most physically extroverted behaviours predominate over sensitive, subjective states of mind. I would speculate, in fact, that it was to some degree precisely because Charcot formulated such a de-sexualized and de-emotionalized model of hysteria in the male that he was able to apply the concept to members of his own sex and that his professional colleagues, after centuries of defensive rejection, were now willing to accept the idea.

Finally, male/female differences may be found in the area of prognostics. A constitutional disorder, hysteria in any patient population, Charcot believed, afforded a gloomy outlook. However, he did hold a degree of hope for the symptomatological control of the disease through specific therapeutic techniques. Hysteria in men, he thought, frequently displayed a greater degree of 'symptomatological fixity' and was therefore more recalcitrant therapeutically.[41] Furthermore, male hysterias often appeared at somewhat later dates in the life of the individual, and they usually manifested fewer symptoms, at times assuming monosymptomatic form. Symptom formations in males also tended to remain unchanged for long periods of time, enduring at times for months, years, or even the remainder of a patient's life. Female hysterias, however, were as a rule capricious and polysymptomatic. 'Singular thing, this male hysteria,' reads a comment of Charcot's from 1887, 'very different in this regard from that of the woman. Seems to be much more serious and to bring with it an infinitely more serious prognosis.'[42] Male/female differences in theory did not always favour men.

CHARCOT, HYSTERIA, AND THE PROCESS OF GENDER RELATIVIZATION

In some ways alike and in other ways different, then – this is the overall picture we form of the Charcotian models. In the final view, however, I believe we should place the accent on the first feature. If we shift our analysis of the hysteria diagnosis from a structural to a longitudinal perspective and consider Charcot's ideas about men and women in the context of hysteria's long and famous history, the most striking feature of this work concerns not the phenomenon of difference but of *sameness*. For millennia of medical history, hysteria, 'la maladie de la matrice,' had been interpreted as a purely female malady. From the Hippocratic texts of the fourth century BC to the writings of mid-nineteenth-century European physicians, hysteria was interpreted as a pathology of femininity. Its diagnosis served as a kind of medical metaphor for everything that male observers found mysterious and unmanageable in the opposite sex. However, in the final quarter of the nineteenth century, the premier theoretician of the disorder offered a full-blown theory of the disorder in the male sex. He devoted 15 years of his career to the scientific investigation of the topic, illustrated his beliefs with dozens of case histories, and advanced his ideas with the full weight of his professional authority. In so doing, he rejected summarily both the genital and characterological interpretations of the disorder which for centuries had provided the theoretical foundations for the most highly gender-differentiated medical systems. In their place, Charcot offered a new and comprehensive theory. As we have seen, this theory continued to contain a significant number of elements that were conspicuously different for male and female patients. But what is perhaps most surprising is how scattered and limited were these gender asymmetries. Furthermore, nineteenth-century definitions of the functional nervous diseases were primarily symptomatological – that is, they were based on descriptive accounts of the external symptoms of the disorders which were lumped together into disease categories. It was above all in this area, in the clinical construction of the diagnosis, that Charcot's ideas about male and female hysteria were extraordinarily alike.

At one time or another, Charcot attributed virtually the entire range of physical behaviours from past conceptualizations of female hysteria to his male patients as well. The sensation of the *globus hystericus* in the throat, which classical medical authors had believed resulted from the pressure of a mobile womb on other organs of the body cavity, appeared regularly in both Charcot's male and female patients. The anaesthesias and hyperaesthesias that Charcot located regularly in his male patients may be seen as secularized and scientized versions of the *stigmati diaboli* of female witches in the sixteenth and seventeenth centuries. And for generations the hysterical attack had been interpreted as a sort of bodily symbolism for childbirth, the female orgasm, and feminine nature in general. During the penultimate decades of the nineteenth century, Charcot described the attack frequently in both the men and women in his clinic.

Perhaps most striking were Charcot's ideas about 'hysterogenic zones.' Hysterogenic zones, it will be recalled, were those small, highly sensitive points or patches on the body which, according to Charcot and his students, could upon stimulation start or stop an hysterical seizure.[43] The distribution of these 'spasmogenous points' in Charcot's male and female patients was different in minor details. Female patients, for instance, presented more points in the mammary area, whereas males revealed greater sensitivity in the shoulders and lower back. More noteworthy, however, was Charcot's attribution to his female patients of hysterical pain in the organ then believed most to differentiate women from men, namely the ovaries. Charcot, we have seen, rejected uncompromisingly the etiological involvement of the genitalia in hysteria; but, as mentioned above, he continued to allow a role for this area of the body in the *symptomatological* realm. In the early 1870s, when assembling his model of feminine hysteria, Charcot had 'discovered' the first hysterogenic point in the ovaries, and went on to claim that hysterical ovaralgias were among the most common indications of the disorder in women.[44] Ten years later, when working with patients of both sexes, he also believed that he located a number of adult men with hysterogenic zones in the area of the lower abdominal wall that corresponded exactly to the position of the ovaries in the female body.[45] Charcot's student Georges Gilles de la Tourette was so impressed by the phenomenon that he labelled these areas 'les zones pseudo-ovariennes' of the male body.[46] Pseudo-ovarian hyperaesthesias, physicians at the Salpêtrière found, were precisely localized and upon compression could produce in these men all the symptoms of the hysterical aura and attack.

The dramatic *rapprochement* between the sexes represented by Charcot's work on hysteria in men and women, then, involved a complicated, two-part process. It was achieved in part through the formulation of separate ideas of etiology, symptomatology, and prognostics. Most often these ideas were in accord with preconceived contemporary notions of male and female natures. At other times, however, Charcot's work involved not a masculinization of the hysteria concept, but the transposition, with remarkable anatomical literalness, of a very old and gynocentric model of sickness on to the bodies of members of the male sex. In another context, I have argued that Charcot's writings on this subject carried a distinct liberalizing potential for science and society. In particular, I suggested that his work contributed to a process of 'gender relativization' whereby the highly polarized sex/gender system of the mid-Victorian period came under criticism and was replaced increasingly by a closer and more flexible set of gender definitions.[47] This process, only implicit in Charcot's work of the 1880s, was advanced much further, I went on to suggest, by Freud a generation later, with his notion of the combined masculine and feminine components in human character and his concept of universal bisexuality. In contemporary feminist psychologies, the theme has been explored in great detail. Interestingly, and rather uncharacteristically, the challenge to the socio-sexual system of the nineteenth century came in this instance from *within* the Victorian medical establishment. With the advantage of a hundred years' hindsight, we can see plainly that Charcot, in his attempt to deal with the hysteria diagnosis equally in men and women, was not as free from the preconceptions of his time as he liked to think. None the less, the mixed results that he achieved, and the sort of stigmatized egalitarianism that he imposed on his male and female patients, represents a significant departure from the dominant theoretical standards of his time.[48]

Finally, there are historiographical implications to one subject. What is perhaps most significant from our analysis is not the success or lack of success, liberalism or conservatism, of a past physician judged by latter-day norms. What is most striking is the discovery of how *complexly and selectively gendered* were Charcot's theories of male and female hysteria. The Charcotian medical model as a whole may be said to be gendered; but, beyond this, the individual components of the theory were gendered in different ways and to different degrees. While some components appear not to have been gendered at all. Each of these features is meaningful historically, and each deserves scholarly study. While the contrasts in Charcot's thinking about hysteria between the sexes are important, it is insufficient to conceptualize men and women in the past only in terms of absolute opposites, and we should guard against a tendency to fetishize the concept of difference/*différence* at the expense of other analytical categories. In the future, we will require a historiographical model of gender, science, and medicine that accounts equally for the differences *and similarities* of men and women in past theoretical systems and for their constantly shifting interrelations in specific historical situations.

CHARCOT, HYSTERIA, AND THE
BRITISH MEDICAL COMMUNITY, 1875–1900

From a different perspective, the themes and issues brought out in an examination of Charcotian theory may be highlighted by considering the reception of these ideas in the medical community in Britain. With the advancing process of professionalization, the final third of the nineteenth century witnessed the formation of a European network of medical journals, conferences, and professional associations. But despite the new and self-proclaimed scientific internationalism of the later part of the century, there endured distinct 'national styles' of medical theorization. The French, German, and British discourses on hysteria illustrate the fact well. Throughout the 1800s, French physicians conducted a long and passionate love affair with the subject, culminating in the *belle époque* of hysteria. After initially receiving Charcot's writings with hostility, German and Austrian physicians also began to embrace the topic in the late 1880s, and during the final decade of the century contributed as much literature on hysteria as the French. In contrast, however, the British medical community throughout this period maintained a long and studied silence.[49]

In the major European Continental medical cultures, as well as in North America, the study of hysteria and the other 'functional nervous disorders' played an important, if not decisive, role in the emergence of both twentieth-century neurology and private-practice psychiatry. Not only Charcot, but Freud, Pierre Janet, and Joseph Babinski devoted ten to fifteen prime years of their careers to the systematic investigation of 'the great neurosis.' The same cannot be claimed for British medical history. The home discipline of Charcot and Freud was clinical neurology, and the final three decades of the nineteenth century, the high point of medical concern with hysteria, was also the golden age of English neurology. But, as W. F. Bynum has pointed out, Charcot's counterparts in the British neurological community showed comparatively little interest in the hysterical neuroses and tended to look with quiet scepticism on the work of their professional brethren across the Channel.[50] The most detailed enquiry in the general area, Charles Handfield Jones's 900-page *Studies on Functional Nervous Disorders*, appeared in the pre-Charcotian period.[51] John Hughlings Jackson, despite an intense interest in the neuropsychology of epilepsy and other convulsive disorders, did not write on hysterical disorders. And David Ferrier limited his researches to cerebral localization. A partial and important exception is William Gowers. In the mid-1880s, Gowers included in his classic *Manual of Diseases of the Nervous System* a highly discerning 40-page discussion of hysteria – all in all, probably the finest piece of medical writing on the subject to come out of Britain in its generation.[52] Gowers, however, presented hysteria as only one of two-dozen functional nervous disorders, and he treated its sister conditions, neurasthenia and hypochondriasis, in short and dismissive sections at the close of the book. Something of a tradition within nineteenth-century British surgery and neurology was the study of 'local hysteria' or the hysterical replication of localized neurological symptoms. This line of research commenced in the 1830s, with the work of Benjamin Brodie on hysterical joint disease, and extended through the century with writings by Hocken, Todd, Althaus, Reynolds, Skey, and Paget. In the closing decades of the century, two books in this tradition were Thomas Buzzard's *On the Simulation of Hysteria by Organic Disease of the Nervous System* (1891) and Henry Charlton Bastian's *Various Forms of Hysterical or Functional Paralysis* (1893). This line of research produced some brilliant clinical observation, and its place in the intellectual history of hysteria remains unappreciated. But, even here, the central scientific attraction in studying the subject was differential diagnostics. Buzzard, Bastian, and their colleagues were interested in hysteria in order to exclude it from organic diagnoses, such as chorea, epilepsy, and disseminated sclerosis, and showed little concern to investigate the phenomenon separately. As Bynum has observed, mainstream British neurology during the late nineteenth and early twentieth centuries increasingly abandoned the study of the 'softer' functional disorders, ceding the subject to the emergent field of office psychiatry, and limiting itself to the investigation of ascertainably organic ailments of the nervous system.[53]

A second medical specialty in Britain that might have been expected to concern itself with hysteria at this time was institutional psychiatry. In the first half of the nineteenth century, the bulk of French medical writing took the form of full-length treatises on hysteria by leading alienists, such as Louyer-Villermay, Landouzy, and Brachet. However, here again we find a relative lack of concern among British physicians during the late nineteenth century. In comparison, for instance, with the ample French and German-language monographic literature from this period on the concept of 'folie hystérique/Irresein hysterische,' there exists only a small body of commentary by asylum doctors in Britain. In an important historical analysis, Michael Clark has argued that British alienists in the second half of the century formed a highly moralized view of non-organic nervous illness. Hysteria, in particular, Clark has suggested, was seen less as a genuine malady, either mental or physical, than as a defect of the will, a deceitful, morbidly egotistical and morally depraved nature. As such, it deserved neither scientific study nor therapeutic ministration.[54]

This general impression of British unconcern with the subject is reinforced by a bibliometric review of titles in the *Index-Catalogue* of the Surgeon-General's Office.[55] Using language and place of publication as indications of nationality, we find for the period 1880–1900 the following numbers of book-length studies of hysteria: France, 133; Germany and Austria, 50; the United States, 31; Italy, 12; England and Scotland, 4; Spain, 3; the Netherlands, 2; and Russia, 1.[56] The heroic age of hysteria, in other words, was centred in those quintessentially *fin de siècle* cities of Paris and Vienna, and did not extend to London and Edinburgh.[57]

During the Charcot era, British doctors tended to look on quizzically at the great fuss being made about hysteria across the Channel. If they commented at all, it was most often to claim that the disease was endemic where studied most closely. Cases of hysteria, they felt, simply were not found, or were found only very exceptionally, among British men and women. Drawing theoretical support from the recent notion of the 'national psychologies,' medical writers in Britain often went on to suggest that hysteria was essentially a disorder of the French, or of Mediterranean peoples in general, or of Slavs or Jews or Arabs. At any rate, not of the solid, sensible, and emotionally-disciplined English.[58] 'That the French as a race are more emotional than either the English or the Germans is a matter of common remark,' reported the Paris correspondent to the *Lancet* in 1890,

and to the casual observer this would seem in keeping with the apparently greater number of hysterical outbursts amongst females which are brought under one's observation in this city. Speaking from an experience of six years' residence, a woman in any form of hysterical fit in the streets of London is a rare occurrence, while a similar exhibition in the streets of Paris is – while short of being an everyday phenomenon – at least not infrequent.[59]

In keeping with this outlook, the first volume of the widely-read *Dictionary of Psychological Medicine*, published in 1892, offered two entries under the heading of hysteria – one, the editor Daniel Hack Tuke explained, 'from the French perspective' and written by Charcot and Pierre Marie, and the other by Horatio Donkin, an internist at Westminster Hospital, 'from the English point of view.'[60] According to Tuke and Donkin, the hystero-epileptic attacks described by Charcot in

France did not occur in Britain.[61] This same medical chauvinism continued into the new century. As late as the eleventh edition of the *Encyclopaedia Britannica*, hysteria was characterized as a matter of 'national idiosyncrasy.' 'Certain races are more liable to the disease than others,' wrote Edwin Bramwell and John Tuke in 1910. 'Thus the Latin races are much more prone to hysteria than are those who come of a Teutonic stock, and in more aggravated and complex forms.'[62] Perusing the medical literature as a whole, the reader forms the sense that hysteria was seen by these doctors as essentially un-English.[63]

If *fin de siècle* physicians in Britain were relatively uninterested in the general theme of hysteria, to the concept of the disorder in males they proved exceptionally resistant. The exploration of hysterical neuroses in men and children, we have learned, was among the most novel and dramatic developments in the European mental sciences during the late nineteenth century, and was perceived as such by Continental observers at the time. But while France witnessed a veritable flood of books and articles on masculine hysteria, and German-language publications came in an increasing stream, a review of the *Index-Catalogue* reveals only a single book-length work by a British author from the last two decades of the century.[64] The leading French and German neurologists in the late eighties and the nineties wrote extensively on male traumatic hysteria. And Continental textbooks of neurology, such as those of Grasset and Oppenheim, registered the debate with discussions of the subject increasing in length with successive editions. But contemporaneous British treatises on nervous diseases, including works by Wilks, Ross, Gowers, Buzzard, and Stewart, virtually ignored the burgeoning Salpêtrian literature on hysteria in men. 'Little or no attempt has . . . been made in this country to organize any collective investigation on this subject,' complained a physician from Edinburgh in 1890.[65] And two years later Donkin still had to admit that masculine hysteria as a topic of scientific enquiry 'has certainly been passed over too lightly by many English writers.'[66] No medical dissertation on male hysteria was written at a British university during this 20-year period. No British medical figure contributed original research to the on-going study of the subject in the late nineteenth century.[67]

British doctors, however, were not entirely silent on the matter. From 1880–1900, in the English, Scottish, Irish, Welsh, and Commonwealth medical presses, I have located 19 articles pertaining to hysterical disorders in the male. Numerous neurological textbooks also include pertinent passages. Nearly all of these articles are short in length and contain only a small number of case histories. In the main, these sources are concerned with the differentiation of organic from hysterical disorders. Moreover, they reveal time and again a range of interpretive devices that served to distance medical men from the reality and significance of widespread hysterical disorders in members of their own sex and country. In the 1840s, the eminent English anatomist Charles Bell had depicted the male hysteric as a classical hero, an Adonis figure in a kind of artful, athletic distress (see plate 7).[68] However, since mid-century, idealization of the disorder became impossible, and in its place appeared a number of alternate means of presentation. An easy approach taken by many writers was simply to establish at the outset the reality but extreme rarity of the disorder in its masculine form. 'The disease, though infinitely more common in females than in males, is not confined to them,' commented one writer typically.[69] And in his *System of Medicine* of 1880, under the section on causes of the disorder, Russell Reynolds began with gender: 'Doubtless the most frequently predisposing cause is that condition of the nervous system which is more or less characteristic of the female sex. Hysteric women are met with daily; hysteric men and boys are of comparatively rare occurrence.'[70] The growing realization on the

Fig. 38. Véritable opisthotonos, d'après Ch. Bell. L'esquisse originale est au collège des chirurgiens d'Édimbourg. «J'ai pris ce dessin sur des soldats blessés à la tête à la bataille de Coronne. Trois hommes étaient semblablement atteints et dans un espace de temps assez court présentèrent les mêmes symptômes, de sorte que le caractère de l'affection ne pouvait être méconnu.» (*The anatomy and philosophy of expression as connected with the fine arts*. By sir Charles Bell. — London, 1872.)

PLATE 7 The Male Hysteric as Classical Hero. Illustration from Sir Charles Bell, *The Anatomy and Philosophy of Expression as connected with the Fine Arts* (1872), reproduced in Paul Richer, *Études cliniques sur la grande hystérie ou l'hystéro-épilepsie*, 2nd edn (1885). Note the absence of genitalia in Bell's representation.

Continent that the disorder might occur in roughly equal numbers of men and women was summarily rejected by British writers.[71] To similar effect, many authors acknowledged the occurrence of the disorder in males in a programmatic phrase or sentence at the beginning of their discussions. They then proceeded to draw their illustrative cases exclusively from females;[72] or to refer throughout their narratives to patients as 'she,' 'her,' or 'the hysterical woman;'[73] or to fall back on old gender-based stereotypes of the disorder.[74]

Other authors offered more subtle interpretive strategies. Many British physicians, for instance, presenting cases that almost certainly would have been labelled hysterical on the Continent, opted for alternate diagnostic terminologies for male patients. 'Local nervous affection,' 'emotional paralysis,' 'paralysis dependent upon idea.' 'nervous mimicry,' 'nervous shock,' 'railway spine,' and 'traumatic neurasthenia' were terms of substitution at once more technical and less pejorative. A related tactic, and a favourite technique in Britain, involved the speculative organic re-diagnosis of cases involving men. A woman complaining of wide-ranging and ill-defined nervous complaints accompanied by dramatic behaviours was judged more or less reflexively to be hysterical. Males, however, presenting identical physical and mental symptomatologies, were much more likely to be viewed as difficult cases of undetermined organic disease.[75] Other practitioners grudgingly applied the label of hysteria to male patients, while remaining tentative about its accuracy and propriety.[76] Still other common strategies included limitation of cited patient materials to boys or adolescents;[77] the description of male hysterics as weak, effeminate, or homosexual men;[78] and the restriction of cases to foreigners.[79] Several articles creatively combined these features. Interestingly, these practices are forms of 'diagnostic camouflage' that have figured prominently in the long medical history of the non-recognition of hysteria in men.[80] They are also, it will be noticed, among the clinical stereotypes against which Charcot argued vociferously.

The unwillingness of British medical figures to take account of contemporary French work on hysteria in the male is revealed further by the history of English-language translations of Charcot's writings. Charcot personally was very fond of Britain. He spoke English fluently and read English literature as a pastime. He was very familiar with nineteenth-century British surgical and neurological literature. He served as an honorary member of the British Medical Association and the Royal Medical and Chirurgical Society of London, and he travelled on several occasions to scientific conferences in the country.[81] Throughout the 1860s and 1870s, medical Britain responded to Charcot with equal enthusiasm and acceptance. One translation after another was undertaken of his books on topics pertaining to internal, geriatric, and neurological medicine. Similarly, Charcot's lectures on hysteria in women from the 1870s were rendered into English promptly upon their appearance in Paris. Under the auspices of the New Sydenham Society, the first volume of the *Leçons sur les maladies du système nerveux*, including six lengthy lectures on female hysterical disorders, was translated by George Sigerson in 1877.[82] Volume 2 of the *Leçons*, also translated by Sigerson, appeared in 1880.[83] Translation of the third volume in the series, however, in which nearly all cases of hysteria occurred in men, languished. This was all the more surprising as Charcot, in formulating his theory of a traumatic male hysterical type, had drawn heavily on the previous medical work of Brodie, Todd, Reynolds, and Paget. Throughout the 1880s, when French interest in hysteria was at its peak, prominent British medical translators continued to offer editions of Charcot's writings, but they limited themselves to his writings from preceding decades which dealt with somatic topics, such as diseases of old age and spinal and cerebral diseases.[84] In 1889, an English edition of the third volume of the *Leçons sur les maladies du système nerveux* finally appeared, translated by Thomas Dixon Savill, a personal enthusiast of Charcot.[85] Remarkably, however, translators took no notice of the four volumes of Charcot's later writings, including the celebrated and brilliant *Leçons du mardi*, appearing in French in 1887 and 1889, and the *Cliniques des maladies du système nerveux*, published in 1892–3, both of which contain significant numbers of case histories of male hysterics. These volumes remain unavailable in English today.[86]

The question of why a given medical subject is *not* studied somewhere is a fine topic of investigation for the historical sociologist of science. British physicians a hundred years ago may well have had sound scientific reasons for responding critically to Charcot's research on hysteria. For generations, Britons had resisted fashionable medical ideas and nostrums from the Continent. The British non-response to Charcot's research in some ways anticipated the more thoroughgoing critique by Hippolyte Bernheim and the Nancy School begun in the late 1880s and the negative posthumous re-evaluation of Charcot's work. Furthermore, the contention of many British doctors that there existed real differences in the clinical manifestation of the disorder should not out of hand be discounted.

According to current-day psychiatric thinking, hysteria represents an attempt to deal with overwhelming psychological anxiety through the unconscious language of the somatic symptom. But the process of psychological socialization differs from culture to culture as well as among individuals, and this includes determination of which forms of stress response are deemed acceptable and unacceptable. If the current view is correct, there may well have existed in the nineteenth century different psychopathological styles – more depressive disorders among the British, and more emotionally extroverted patterns of neurosis in other societies.

However, I suspect that all in all the contrasts in national patterns of theorization about hysteria a century ago had less to do with objective differences in the clinical picture presented by patients than with the social, cultural, and psychological attitudes of doctors. The two decades of cultivated neglect of male hysteria in British medical circles, the attempt of physicians to characterize the neurosis as a foreign pathology, and their efforts to limit the label to stereotyped or stigmatized populations are not without historical meaning. From Robert Burton's *Anatomy of Melancholia* of 1621, through the work of Thomas Willis and Thomas Syndenham in the 1680s, to the book-length treatises of Bernard de Mandeville, Richard Blackmore, Nicholas Robinson, Robert Whytt, and many others, English and Scottish writers had theorized richly about the 'nervous distempers.' With a mixture of pride and resignation, they had accepted George Cheyne's account in the 1730s of the nervous disorders as 'the national malady.' And throughout the seventeenth and eighteenth centuries, they had discussed a host of neurotic conditions – nervousness and nostalgia, the spleen and the vapours, hysteria, melancholia, hypochondriasis, and lycanthropy – fully and comfortably in members of both sexes.[87] With the nineteenth century, however, things changed. In an age of medical materialism, and within the highly moralized context of Victorian culture, chronic nervous disease ceased to connote a refined sensibility or Romantic eccentricity. Rather, it now implied a form of physical and mental degeneracy and a crude and uncontrolled emotionality.

For middle-class male diagnosticians then, these associations were especially unacceptable in regard to individuals of their own sex and class. Societies, we know, vary widely in their normative definitions of masculinity and femininity in both sickness and health. They also differ significantly in the degree of explicitness with which these gender roles are discussed in public discourse and in the amount of overlapping, crossgenderal behaviour to be tolerated. In this light, it is perhaps not surprising that nineteenth-century doctors were more open to Charcot's exploration of male hysteria in *fin de siècle* Paris and Vienna than in either late Victorian and Edwardian Britain or Wilhelminian Germany. The ideology of separate spheres, strongly buttressed by the medical establishment, was nowhere stronger than in Britain during the final third of the last century. The new Continental literature on hysteria in women could with relative ease be assimilated into traditional medical ways of thinking about both the disease and feminine nature. However, the theme of masculine hysteria was altogether different. Charcot's hysterization of the male body in the 1880s was sharply at variance with dominant medical models of masculinity, and it ran counter to reigning Victorian codes of manliness. Moreover, it required from Victorian physicians the application of an ancient and denigratory label to members of their own sex. And perhaps most disturbing, it suggested the possibility of exploring the feminine component in the male character itself. With its implicit relativizing of gender identities, a detailed and comparative study of hysterical men and women was a gesture profoundly uncongenial to middle-class male diagnosticians in Britain. With impressive consistency, therefore, they either argued the idea away or ignored it altogether. Gender relativization, it appears, had occurred at different times and different rates from country to country.

Finally, considerable evidence exists to suggest that the process has also taken place differently within the medical specialties. I have said that there was little commentary on hysteria in the late nineteenth century from British alienists, neurologists, and general practitioners. But one professional group commented steadily throughout this period on the topic. Until the turn of the century, British gynaecologists continued to write on hysterical neuroses with authority and self-confidence as if the subject remained fully within their domain of expertise. In the 1830s and 1840s, Thomas Addison and Thomas Laycock respectively had initiated a tradition, which would endure in Britain for three generations, of studying hysteria in the format of the general gynaecological treatise. The writings of Addison and Laycock united Hippocratic ideas of uterine pathology, the neurological literalism of seventeenth-century English and Scottish authors, and the new reflex physiology of Marshall Hall. It proved a powerful but pernicious synthesis. According to Addison and Laycock, a dauntingly wide range of mental and physical ailments in women traced to derangements of the reproductive system. Hysteria in particular, they believed, resulted from 'uterine irritation' which, operating through the highly intricate sympathetic nervous system, could affect all other body organs, including the brain. The generative organs of women, above all the uterus and the surrounding network of nerves, and female reproductive physiology in general, especially irregularities of menstruation, parturition, and lactation, were the primary causes of the disorder; the control of these organs and regulation of the processes were its remedies.[88] Through the mid-century period, this refurbished uterine theory remained influential.[89] And to this gynaecological model was added a characterological conception of the disorder when Robert Brudenell Carter in 1853 published *On the Pathology and Treatment of Hysteria*.[90]

In the final third of the century, this heritage of theories came into full play in the literature of British gynaecology. In France, Charcot had succeeded in silencing, more or less permanently, the gynaecological claim to authority over hysteria. In the Anglo-American medical world, however, no figure of comparable professional prestige existed, and, as a consequence, psychological, neurological, and gynaecological models competed for prominence.[91] The most influential English author on hysteria and neurasthenia throughout this period was William S. Playfair. Society doctor, physician to the Duchess of Edinburgh, and Professor of Obstetrical Medicine at King's College, Cambridge, Playfair wrote frequently about the nervous disorders. While he did not speculate openly on matters of causation or pathology, Playfair's symptomatological observations as well as the remedies he prescribed reveal his etiological assumptions. In his case histories, he carefully noted urogenital symptoms. He correlated the appearance of hysterical symptoms with events in the marital and maternal lives of his patient. And he at times prescribed the rest-cure of Silas Weir Mitchell (of which he was the major advocate in the country) in tandem with local gynaecological treatments.[92]

Similarly, the leading academic physician writing on hysteria in Britain at this time was one Graily Hewitt, Professor of Midwifery and Gynaecology at University College and sometime President of the Obstetrical Society of London. Simply put, Hewitt was an unreconstructed uterinist. He believed that hysteria was caused by the direct reflex excitability of the uterus. But Hewitt also formulated an elaborate theory of his own according to which hysteria resulted from the 'anteflexion' or 'retroflexion' of the uterus, that is, from a bending of the organ at the center, which, in turn, led to the 'traumatic congestion' of the blood supply through the organ. At the International Medical Congress of 1881, in London, Hewitt demonstrated this morbid action with the use of a sponge model of the uterus six times life size. In the fourth edition of his textbook the following year, he officially presented his new theory, illustrated with some 18 cases, as a major medical breakthrough.[93]

The ideas of Playfair and Hewitt were shared by a substantial majority of their professional peers during the final three decades of the century. According to the British gynaecological literature, a much larger portion of the bodies of women is given over to reproductive physiology than in men, and this delicate machinery is continually prone to malfunction. As the nerves of the pelvic viscera disperse throughout the body, these organs are capable of causing all sorts of nervous havoc. The key physiological concept for these authors was the 'genital reflex neurosis,' and their medical motto, *Le bassin c'est la femme.*[94] Precisely which part of the female pelvic topography was at fault remained a matter of debate. The uterus proper, the surrounding plexus of nerves, the fallopian tubes, and the ovaries were at one time or another implicated. This doctrine of gynaecological determinism was often accompanied by a series of supporting ideas. The age of susceptibility to hysteria was restricted to the period from puberty to menopause. The disorder was believed to be greatly exacerbated by menstruation, and celibacy, masturbation, and sexual over-indulgence could serve as direct causes. The wide-ranging physical manifestations of the disorder, we also learn here, were frequently joined by vagaries of behaviour. The hysterical woman might be duplicitous, self-dramatizing, and morally undisciplined – in a word, she had the 'hysterical disposition.' Furthermore, these characteristics were often interpreted as a natural and inevitable part of the female condition.[95] Throughout this writing, the reproductive organs are seen to exert a controlling influence over women. And the entire literature is based on the idea of the innate and inferior nervous and mental constitution of woman.[96] Moreover, a number of the most prominent institutional alienists of the age continued to associate hysterical sickness with nymphomania and erotomania and to classify the disease as a form of 'ovarian insanity' and 'uterine insanity.' Despite an occasional outspoken dissenter, these beliefs prevailed in the British gynaecological profession until the century's end.[97]

The therapeutic implications of these views ranged from the absurd to the sinister. A significant number of British gynaecologists – they tended to be those with lucrative private practices – followed Playfair in emphasizing the rest-cure for their hysterical female patients. However, many others, mainly doctors at large hospitals, public infirmaries, and out-patient clinics, deployed a raft of localized physicalistic procedures. In the doctor's office, a gynaecological theory of hysteria required not neurological tests or psychological observation, but a vaginal examination. Use of the speculum seems to have been commonplace procedure, including with unmarried and adolescent patients. One reads endlessly in the literature of douches and dilations, intra-uterine injections, and gynaecological electrotherapeutics. The insertion of 'stems' and pessaries to restore the uterus to its natural shape was also very frequent. For persistent symptoms, there was the administration of chloral and moraphine. And a number of authors advised marriage and pregnancy. Intense ovarian pressure was often employed too. For severe and chronic cases more drastic measures were endorsed. A number of texts advocated the vulvar application of leeches, blistering of the clitoris, and cervical and clitoral cauterizations. Some practitioners developed their own prescriptions. Hewitt, for instance, proposed 'chloroform inhalations,' 'temporary suffocation of the patient,' and 'injection of iced water into the rectum.'[98] Numerous physicians advocated, and practised, unilateral and bilateral oophorectomy (removal of the ovaries) for hysterical disorders.[99]

CONCLUSION

Needless to say, the idea of hysteria in the male plays no part whatsoever in the literature of nineteenth-century British gynaecology. The citation of Charcot's writings in this writing was highly selective. Gynaecological physicians referred to Charcot's work in order to document the neurosymptomatology of a case, and they cited with approbation his notions of iliac hyperaesthesias and ovarian hysterogenic points. But the medical commentary on hysterical disorders in adult men that was then proliferating in France and Germany does not appear in these pages. Several writers in this camp specifically argued against the concept of male hysteria.[100] And a number of them openly privileged the Hippocratic texts over modern theories of the disease.[101] Mainly, there was no recognition of the subject. It was as if Charcot had never written.

Within the period of the modern scientific world view, with its naturalistic conceptualization of health and disease, the recognition that men and women could suffer equally from debilitating nervous disorders appeared first in England and Scotland. This occurred during the seventeenth and eighteenth centuries in medical and literary sources. The first half of the nineteenth century, however, brought the organization of medicine into a professional male collectivity and the development of a new complex of social, sexual, and moral attitudes. By all indications, the notion of hysteria in men was then officially 'forgotten' by European physicians and remained so for two generations. The idea was then re-learned, first in France, during the intellectually liberal years of the early Third Republic, in the writings of Charcot. The realization came to Germany and Austria, in part through the work of Freud, a decade later. However, throughout hysteria's *belle époque*, a majority of members of the British medical community retained a belief in the dual doctrines of gynaecological determinism and an innate hysterical constitution. As a result, they remained resolutely closed to the concept. In Britain, the idea was not achieved again until the experience of the First World War, with its inescapable epidemics of 'shell-shock' and with the subsequent, psychoanalytically-informed literature on the war neuroses in soldiers.[102] Not coincidentally, the rise and fall of the British gynaecological paradigm of hysteria, with its implicit but programmatic denial of the phenomenon in men, corresponds closely in chronology with the rise and fall of the Victorian socio-sexual system as a whole.

ACKNOWLEDGEMENTS

The research for this essay was supported by a grant from the Wellcome Foundation through the London Unit of the Wellcome Institute for the History of Medicine. The essay was delivered first in lecture form on April 6, 1990, at a symposium on the History of Hysteria at the National Hospital, Queen Square, London. I would like to thank the members of that audience for their comments; Marina Benjamin for her patience and encouragement; and Ruth Harris for alerting me to the interesting subject of Charcot's reception in the British medical community.

NOTES

1 For an exception to the rule, see Elaine Showalter, *The Female Malady: Women, Madness, and English Culture, 1830–1980* (Pantheon, New York, 1985), ch. 7.

2 Harry Brod (ed.), *The Making of Masculinities: The New Men's Studies* (Allen and Unwin, Boston, 1987), pp. 1–17.

3 Joan W. Scott, 'Gender: a useful category of historical analysis,' *American Historical Review*, 91 (1986), pp. 1053–75, at p. 1054; Natalie Zemon Davis, 'Women's history in transition: the European case,' *Femininist Studies*, 3 (1975–6), p. 90.

4 For intellectual histories of the phenomenon, see Ilza Veith, *Hysteria: The History of a Disease* (University of Chicago, Chicago, 1965); George R. Wesley, *A History of Hysteria* (University Press of America, Washington, DC, 1979); and Étienne Trillat, *Histoire de l'hystérie* (Seghers, Paris, 1986).

5 For a detailed review of the burgeoning scholarly literatures on hysteria, consult Mark S. Micale, 'Hysteria and its historiography: a review of past and present writings,' 2 parts, *History of Science*, 27 (1989), pp. 223–61; 319–51. An attempt to establish a set of interpretive guidelines for the prospective study of the subject may be found in Micale, 'Hysteria and its historiography – the future perspective,' *History of Psychiatry*, 1 (1990), pp. 33–124.

6 Howard M. Feinstein, 'The use and abuse of illness in the James family circle: a view of neurasthenia as a social phenomenon' in Robert J. Brugger (ed.), *Our Selves/Our Past: Psychological Approaches to American History* (Johns Hopkins University Press, Baltimore and London, 1981), pp. 228–9.

7 S. Weir Mitchell, 'The evolution of the rest treatment,' *Journal of Nervous and Mental Disease*, 31 (1904), pp. 368–73.

8 F. G. Gosling, *Before Freud: Neurasthenia and the American Medical Community, 1870–1910* (University of Illinois Press, Chicago, 1987), p. 34.

9 Edward Shorter, 'Women and Jews in a private nervous clinic in late nineteenth-century Vienna,' *Medical History*, 33 (1989), pp. 149–83, esp. p. 174.

10 Showalter, *Female Malady*, chapter 7; Esther Fischer-Homberger, *Die traumatische Neurose: vom somatischen zum sozialen Leiden* (Hans Huber, Bern, 1975), pp. 105–70; Martin Stone, 'Shellshock and the psychologists' in W. F. Bynum, Roy Porter, and Michael Shepherd (eds), *Anatomy of Madness: Essays in the History of Psychiatry*, (3 vols, Tavistock, London and New York, 1985), 2, 242–71.

11 A full study of the subject may be found in Mark S. Micale, 'Diagnostic Discriminations: Jean-Martin Charcot and the nineteenth-century idea of hysteria in the male (Doctoral Dissertation, Yale University, 1987); and 'Charcot and the idea of hysteria in the male: a study of gender, mental science, and medical diagnostics in late nineteenth-century France,' *Medical History*, 34 (1990), pp. 363–411.

12 Jean-Martin Charcot, *Leçons sur l'hystérie virile*, introduction by Michèle Ouerd (SFIED, Paris, 1984), p. 11.

13 Jean-Martin Charcot, *Leçons sur les maladies du système nerveux* (henceforth *LMSN*), ed. Bourneville (1872–8) in *Oeuvres complètes de J.-M. Charcot* (9 vols, Bureaux du Progrès Médical, Delahaye and Lecrosnier, Paris, 1886–93), 1 (1892), pp. 275–405, 427–48.

14 Micale, 'Charcot and the idea of hysteria in the male.'

15 Charcot, 'De l'hystérie chez les jeunes garçons,' 2 parts, *Le progrès médical*, 10 (1882), pp. 985–7, 1003–4.

16 This figure includes cases in the 3 vols of *Leçons sur les maladies du système nerveux*, the 2-volume *Leçons du mardi*, and the 2 vols of *Clinique des maladies du système nerveux* as well as in various medical periodicals. It should be noted that I am not including in my analysis the *Iconographie photographique de la Salpêtrière* . While fascinating and important, it is my view that this well-known set of documents has been given disproportionate weight in recent historical studies of Charcot. The relationship of these visual materials on hysteria, assembled between 1877 and 1880 by two of Charcot's students, to Charcot's own detailed clinical writings is complex and problematic. In the present essay, my reading will rest on the case-historical writings of Charcot which, in form and authorship, represent congruent source materials for the theories of male and female hysteria. Readers with an interest in the iconography of hysteria may consult Georges Didi-Huberman's study, *Invention de l'hystérie: Charcot et l'iconographie photographique de la Salpêtrière* (Macula, Paris, 1982).

17 Charcot, *LMSN*, 3 (1890), lecture 6, p. 89; *LMSN*, 3, lecture 18, pp. 253–4; Charcot, *Leçons du mardi à la Salpêtrière . Professeur Charcot. Policliniques*, (2 vols, Bureaux du Progrès Médical, Paris, 1887–9), 2, lesson 5, p. 96; and Charcot and Paul Richer, *Les démoniaques dans l'art* (Delahaye and Lecrosnier, Paris, 1887), pp. 96–106. This point was also emphasized by Georges Guinon in 'L'hystérie chez l'homme comparée à l'hystérie chez la femme,' *Gazette médicale de Paris*, 56 (1885), pp. 231–4.

18 Charcot, *LMSN*, 3, lectures 7 and 8, pp. 97–123.

19 On the resurgence of sexual theories of nervous and mental disease in the nineteenth century, see Mark D. Altschule, 'Venus Ascendant' in *Roots of Modern Psychiatry: Essays in the History of Psychiatry*, 2nd edn rev. (Grune and Stratton, New York and London, 1965), pp. 101–18.

20 Théodore Tarczylo has determined that the volume of anti-masturbation literature in Europe and America peaked during the 1880s and 1890s: *Sexe et liberté au siècle des Lumières* (Presses de la Renaissance, Paris, 1983), appendix 5, pp. 297–8.

21 Showalter, *Female Malady*, pp. 167–8.

22 Augustin Fabre, *L'hystérie viscérale – nouveau fragments de clinique médicale* (A. Delahaye and E. Lecrosnier, Paris, 1883), p. 3.

23 The fullest statement, from a student of Charcot, is Jules Déjerine, *L'hérédité dans les maladies du système nerveux* (Asselin and Houzeau, Paris, 1886), with remarks concerning hysteria on pp. 116–25.

24 Thus, studies such as Jacques Roubinovitch's 'Hystérie mâle et dégénérescence' (Doctoral Dissertation, Paris Medical Faculty, 1890).

25 An amusing statement of this view in the Anglo-American literature was Robert Thornton's *The Hysterical Woman: Trials, Tears, Tricks and Tantrums* (Donohue and Henneberry, Chicago, 1893).

26 Charles Lasègue, 'Les hystériques, leur perversité, leurs mensonges,' *Annales médico-psychologiques*, ser. 6, 6 (1881), pp. 111–18; Henri Huchard, 'Caractère, moeurs, état mental des hystériques,' *Archives de neurologie*, 3 (1882), pp. 187–211; Henri Legrand du Saulle, *Les hystériques: état physiques et état mental. Actes insolites, délicteuex et criminels* (J. B. Baillière, Paris, 1883), pp. 581–625.

27 Charcot, 'De l'hyperesthésie ovarienne', *LMSN*, 1, Lecture II, pp. 283–304; *Leçons du mardi*, 2, lesson 17, p. 402; Charcot and Richer, 'Hysteria mainly hystero-epilepsy', *Dictionary of Psychological Medicine*, 1, p. 637.

28 Charcot, 'Deux cas de contracture hystérique d'origine traumatique,' *LMSN*, 3, lecture 8, p. 117; Case of Antoine Charnu, Bibliothèque Charcot; Case of Jean-Pierre Mattivet, Bibliothèque Charcot.

29 Charcot, 'Deux cas de contracture hystérique d'origine traumatique,' *LMSN*, 3, lecture 8, p. 115.

30 Charcot, *Leçons du mardi*, 2, lesson 12, p. 261. See also p. 256. Peter N. Stearns has discussed the great perceived differences between lower- and middle-class masculinity during the nineteenth century in *Be a Man! Males in Modern Society* (Holmes and Meier, New York and London, 1979), chs. 4 and 5.

31 Charcot's clinical notes at the Bibliothèque Charcot today include many observations of effeminacy in his male hysterical patients. Of one youngster from the Hérault, he recorded that the patient displayed 'féminisme – il joue de petites filles,' and of another he commented that the patient was 'maigre et délicat, les muscles sont faibles, la peau est blanche.' In 1883 the alienist Henri Legrand du Saulle described a visit by Charcot to a young adolescent male demonstrating a number of hysterical signs: 'M. Charcot est appelé en consultation. Il remarque de suite le caractère particulièrement féminin de l'enfant qui portait une bague au doigt, aimait à se parer et jouer à des jeux de petit fille': Le Grand du Saulle, *Les hystériques*, p. 19.

32 Pierre Briquet, *Traité clinique et thérapeutique de l'hystérie* (J. B. Baillière, Paris, 1859). p. 36; Charcot, *LMSN*, 3, lecture 8, p. 114.

33 Charcot, 'Deux cas de contracture hystérique d'origine traumatique,'

LMSN, 3, lecture 8, p. 115; *Leçons du mardi*, 1, lesson 11, p. 208.

34 Charcot, *LMSN*, 3, lecture 8, pp. 117–23; *LMSN*, 3, lecture 18, pp. 257–63; *Leçons du mardi*, 2, lesson 13, pp. 285–92.

35 A concise review of the Salpêtrian symptomatology of hysteria may be found in Charcot and Pierre Marie, 'Hysteria, mainly hystero-epilepsy' in Daniel Hack Tuke (ed.), *A Dictionary of Psychological Medicine* (2 vols, J. and A. Churchill, London 1892), 1, pp. 629–39.

36 This practice was especially common in later cases. See *Leçons du mardi*, 1, lesson 4, pp. 62–5; 2, lesson 2, pp. 19–37; lesson 7, pp. 131–9; lesson 12, pp. 261–5; lesson 13, pp. 292–9; appendix 1, pp. 528–35.

37 *Leçons du mardi*, 2, lesson 6, pp. 124–5; *Leçons du mardi*, 2, lesson 19, pp. 456, 457.

38 Charcot, *Leçons du mardi*, 2, lesson 3, p. 50.

39 For the hysterical attack in female patients, see Charcot, 'Descriptions de la grande attaque hystérique,' *Progrès médical*, 7 (1879), pp. 17–20; 'De l'hystéro-épilepsie' *LMSN*, 1, lecture 13, pp. 321–37; and 'Hystérie à grandes attaques' in *Leçons du mardi*, 1, pp. 103–5. For males, consult the 6 cases in lectures 18 and 19 of the *Leçons sur les maladies du système nerveux*, 3, pp. 253–98.

40 For hysterical attacks in males lacking the third passional period, see the cases of 'Gui' (*LMSN*, 3, lecture 18, pp. 278, 281); 'Laf . . . cque' (*Leçons du mardi*, 2, lesson 12, pp. 265–9); 'Lap . . . sonne' (*Leçons du mardi*, 2, lesson 17, pp. 393–9); and 'P . . . eyn' (*Leçons du mardi*, 2, p. 21, pp. 518–23).

41 Charcot, *Leçons du mardi*, 2, appendix 1, p. 533.

42 Charcot, 'Des paralysies hystéro-traumatiques chez l'homme,' *Semaine médicale*, 7 (1887), p. 491.

43 Charcot, 'Des zones hystérogènes,' *Progrès médical*, 8 (1880), pp. 1036–8.

44 Charcot, 'De l'hémianesthésie hystérique,' *LMSN*, 1, lecture 10, pp. 301–3; 'De l'hyperesthésie ovarienne,' *LMSN*, 1, lecture 11, pp. 320–46.

45 Charcot, *Leçons du mardi*, 2, lesson 19, pp. 454, 456; 'Des paralysies hystéro-traumatiques chez l'homme,' *Semaine médicale* (1877), p. 491.

46 Georges Gilles de la Tourette, *Traité clinique et thérapeutique de l'hystérie d'après l'enseignement de la Salpêtrière* , (3 vols, Plon, Nourrit and Cie, Paris, 1891), 1, pp. 299–300.

47 Micale, 'Charcot and the Idea of Hysteria in the Male,' pp. 409–11.

48 The phrase 'stigmatized egalitarianism' was suggested to me by Roy Porter.

49 The intellectual history of hysteria in Britain during the nineteenth and twentieth centuries remains almost entirely unwritten. Robert Brudenell Carter's *On the Pathology and Treatment of Hysteria* (1853) has of late attracted scholarly attention; but this is an exception, and the major historical surveys of the subject (Veith, Trillat, and Wesley) do not consider the subject. The comparative European reception of Charcot's work is also

an interesting, important, and largely unexplored topic. Some remarks on Charcot and the British medical community may be found in William Parry-Jones, '"Caesar of the Salpêtrière": J.-M. Charcot's impact on psychological medicine in the 1880s,' *Bulletin of the Royal College of Psychiatrists*, 11 (1987), pp. 150–3. Hannah S. Decker discusses the German-language writing on hysteria in *Freud in Germany: Revolution and Reaction in Science, 1893–1907* in *Psychological Issues*, 11, monograph 41 (International Universities Press, New York, 1977), ch. 2.

50 W. F. Bynum, 'The nervous patient in eighteenth- and nineteenth-century Britain: the psychiatric origins of British neurology' in Bynum, Porter, and Shepherd (eds), *Anatomy of Madness*, 1, pp. 89–102.

51 C. Handfield Jones, *Studies on Functional Nervous Disorders* (Churchill, London, 1870).

52 W. R. Gowers, *A Manual of Diseases of the Nervous System*, 1st edn (2 vols, J. and A. Churchill, London 1888), 2, pp. 903–46; 2nd edn (2 vols, J. and A. Churchill, London, 1893), 2, pp. 984–1030. Gowers included in his chapter many references to the latest Salpêtrian literature and illustrated his text with the drawings of Paul Richer.

53 Bynum, 'The nervous patient,' pp. 96–9.

54 Michael J. Clark, 'The rejection of psychological approaches to mental disorders in late nineteenth-century British psychiatry' in Andrew Scull (ed.), *Madhouses, Mad-doctors, and Madmen: The Social History of Psychiatry in the Victorian Era* (University of Pennsylvania Press, Philadelphia, 1981), pp. 271–312 and ' "Morbid introspection," unsoundness of mind, and British psychological medicine, *c*.1830–*c*.1900' in Bynum, Porter, and Shepherd (eds), *Anatomy of Madness*, 3, pp. 71–101. Clark bases his interpretation on the comments of figures such as Henry Maudsley, William Carpenter, George Savage, Charles Mercer, Thomas Clouston, Horatio Donkin, and Joseph Ormerod, drawn from psychiatric textbooks and medical dictionaries.

55 *Index-Catalogue of the Library of the Surgeon-General's Office, United States Army*, 1st series (Government Printing Office, Washington, 1885), 6, pp. 750–67; 2nd series, (1902), 7, pp. 772–804. These figures include medical dissertations.

56 The four British studies were Graily Hewitt, *The Exciting Cause of Attacks of Hysteria and Hystero-Epilepsy* (London, 1881); Thomas Buzzard, *On the Simulation of Hysteria by Organic Disease of the Nervous System* (Churchill, London, 1891); Henry Charlton Bastian, *Various Forms of Hysterical or Functional Paralysis* (Lewis, London, 1893); and Thomas G. Stewart, *Lectures on Giddiness and on Hysteria in the Male*, 2nd edn (Y. J. Pentland, Edinburgh, 1898).

57 Simon Wesseley is finding a similar lack of interest among professional

Britons in the contemporaneous concept of neurasthenia in *The Rise and Fall of Neurasthenia* (work in progress).

58 In response, French physicians insisted defensively on the universality of the disease. See Charcot, *Leçons du mardi*, 2, lesson 18, p. 427; Dr Glorieux, 'L'hystérie chez l'homme,' *Archives médicales belges*, 31 (1887), p. 234; and Alexandre Souques, 'De l'hystérie mâle dans un service hospitalier,' *Archives générales de médecine*, 2 (1890), p. 170.

59 N. A., 'Treatment of Hysterical Fits,' *Lancet*, 2 (November 15, 1890), p. 1060.

60 Horatio Donkin, 'Hysteria' in Tuke (ed.), *Dictionary of Psychological Medicine*, 1, pp. 618–27; and Charcot and Marie, 'Hysteria mainly hystero-epilepsy,' *Dictionary of Psychological Medicine*, pp. 627–41: the quotations appear at p. 618n.

61 Interesting in the light of my previous analysis was the related observation of Gowers that the phase of *attitudes passionnelles* in particular was absent from the convulsions of British patients. See Gowers, *Manual of Diseases of the Nervous System*, 2nd edn, p. 1009.

62 J. B. T. [John Batty Tuke] and E. Bra. [Edwin Bramwell], 'Hysteria,' *Encyclopaedia Britannica*, 11th edn, (29 vols, Enclyclopaedia Britannica Company, New York, 1910), 14, p. 211.

63 Physicians from other countries occasionally responded sharply to this attitude: 'Hysteria can occur in either sex, at any age, and in every race and country. The assertions made by some writers, notably English and German, that hysteria does not occur in their respective countries in all its classical phases, are erroneous . . . Such assertions . . . are evidences either of defective observation or of a prejudice respecting hysteria that is not more enlightened than the ancient belief that the disease has its seat in the womb, and is confined, therefore, to the female sex': James Hendrie Lloyd, 'Hysteria' in Francis X. Dercum (ed.), *A Text-book on Nervous Diseases by American Authors* (Lea Brothers, Philadelphia, 1895), p. 95.

64 Stewart, *Lectures on Giddiness*: Stewart's book, it should be added, consists of four lectures, running to 89 pages in length, and includes only a single case history of a male hysteric.

65 A. Stodart Walker, 'Some notes on "hysteria," with special reference to "hysteria" in the male,' *Edinburgh Medical Journal*, 40 part 1 (1894), pp. 312–22, at p. 313.

66 Donkin, 'Hysteria,' p. 624. See the similar complaint by J. Dreschfeld in 'On hysteria in the male coming on after an injury,' *The Medical Chronicle*, 5 (December, 1886), p. 169.

67 To the best of my knowledge, there exists only one partial exception to the rule. John Michell Clarke, an internist and pathologist at the Bristol General Hospital, wrote extensively and intelligently on hysterical disorders

during the 1890s. Clarke's essays reveal complete conversance with the contemporary French and German medical literatures on the subject. In two long and detailed bibliographical essays in *Brain*, Britain's leading neurological journal, Clarke attempted to bring the burgeoning Continental writing to his countrymen. See 'On hysteria,' *Brain*, 15 (1892), pp. 522–612 and 'Hysteria and neurasthenia,' *Brain*, 17 (1894), pp. 119–78, 263–321. (These essays were later united in Clarke, *Hysteria and Neurasthenia* [Lane, London, 1905]). Clarke's writings are notable for a rejection of racial, characterological, and gynaecological explanations of hysteria, full acceptance of the idea of hysteria in both sexes (Clarke estimated that the disorder occurred between the sexes in a 1:4 ratio), and a strong awareness of the latest psychological interpretations of Möbius, Strümpell, Janet, Freud, and Breuer. Clarke also had a particular interest in the masculine forms of the malady. In addition to his discussion of the subject in the *Brain* article of 1892, see 'Hysteria in man,' *Journal of Mental Science*, 33 (1887–8), pp. 543–6 and 'On three cases of hysteria in men,' *Brain*, 14 (1891), pp. 523–37. Clarke's goal in these writings, it should be added was mainly to educate the British medical readership in the Continental literature rather than contribute to the research himself.

68 Bell's illustration first appeared in the 3rd enlarged edition of *The Anatomy and Philosophy of Expression as Connected with the Fine Arts* (John Murray, London, 1844), p. 160.

69 George Johnson, 'A lecture on hysteria' in *Medical Lectures and Essays* (London, J. and A. Churchill, London, 1887), pp. 266–85, at p. 267.

70 J. Russell Reynolds, 'Hysteria' in J. Russell Reynolds (ed.), *A System of Medicine* (3 vols, Henry C. Lea, Philadelphia, 1880), pp. 630–49, at p. 631. See also James Oliver, 'A case of hystero-epilepsy in the male,' *Brain*, 8 (1885), p. 397; *Fagge's Principles and Practice of Medicine*, 2nd edn, compiled from the manuscript of Charles H. Fagge by Philip H. Pye-Smith (2 vols, Churchill, London, 1888), 1, p. 842; Thomas D. Savill, 'Case of hysteria minor and "ovarian phenomena" in a male subject,' *Lancet*, 1 (May 11, 1889), pp. 934, 935; and Francis Bisshopp, 'A Case of hystero-epilepsy in the male; right-hemiplegia, accompanied with analgesia and anaesthesia,' *Lancet*, 2 (July 27, 1889), pp. 163–4.

71 See, for instance, Walker, 'Some notes on "hysteria"', p. 313.

72 Samuel Wilks, *Lectures on Diseases of the Nervous System* (J. and A. Churchill, London, 1878), pp. 361–80.

73 Julius Althaus, *On Epilepsy, Hysteria, and Ataxy: Three Lectures* (John Churchill, London, 1866), pp. 31–73; Johnson, 'Lecture on hysteria,' pp. 266–85; Donkin, 'Hysteria,' pp. 618–27; Gowers, *Manual of Diseases of the Nervous System*, 2nd edn, pp. 984–1030.

74 Donkin, 'Hysteria,' pp. 618–27.

75 A particularly clear example of this procedure is John S. Bristowe, 'Hysteria and its counterfeit presentments,' *Lancet*, 1 (June 13–20, 1885), pp. 1069–72; 1113–17. Bristowe offers a series of studies in differential diagnosis in which all the male cases are judged to be organic illnesses (anaemia, epilepsy, tuberculosis, and hypertrophy of the oesophagus) while those involving women retain their hysterical status.

76 L. Russell, 'Case of "hysterical" retention of urine in a man,' *Medical Times and Gazette*, 1 (April, 1882), p. 353; M. W. H. Russell, 'A case of hysterical(?) rapid breathing in a youth', *Brain*, 6 (1883), pp. 378–81; Walker, 'Some Notes on "Hysteria".'

77 Wilks, *Lectures on Diseases of the Nervous System*, pp. 385–88; Dr Ingle, 'Hysteria in a Boy,' *Lancet*, 2 (July 21, 1883), p. 106; James Ross, *A Treatise on the Diseases of the Nervous System*, 2nd edn rev. (2 vols, J. and A. Churchill, London, 1883), 2, pp. 859–60; Thomas Savage, 'Case of marked hysteria in a boy of eleven years,' *Journal of Mental Science*, 31 (1885–6), p. 201; David B. Lees, 'Two Cases of Hysteria in Boys,' *Lancet*, 1 (June 9, 1888), pp. 1123–6; William R. Gowers, *Epilepsy and other Chronic Convulsive Disorders*, 2nd edn (J. and A. Churchill, London, 1901), pp. 183–7.

78 Reynolds, 'Hysteria,' p. 631; *Fagge's Principles and Practice of Medicine*, p. 837; Bisshopp, 'A case of hystero-epilepsy in the male,' pp. 163–4; Gowers, *Manual of Diseases of the Nervous System*, 2nd end, pp. 985, 1019 and *Epilepsy and other Convulsive Disorders*, p. 187.

79 Althaus, *On Epilepsy, Hysteria, and Ataxy*, pp. 41–3; C. Shah, 'An interesting case of hysteria in a boy,' *Indian Medical Record*, 23 (1888), p. 302; Theodore Diller, 'A case of hysteria in a boy, characterized by regularly recurring attacks of lethargy: treatment by hypnotism,' *Brain*, 16 (1893), pp. 556–61; Joseph Benjamin, 'Strange hysterical fits in a young man,' *Indian Medical Record*, 6 (1894), p. 144; J. R. M. Thomson, 'A case of hysteria in a male,' *Intercolonial Medical Journal of Australasia*, 2 (1897), pp. 660–3.

80 The concept of 'diagnostic camouflage' has been developed by Lucien Isräel in *L'hystérique, le sexe et le médecin* (Masson, Paris, 1979), pp. 60–2.

81 As Thomas Clifford Allbutt, Regius Professor of Physics at Cambridge reflected, 'No Continental physician was ever more cordially esteemed by Englishmen – not even Trousseau – because none had been more open to English ideas or more familiar with English work': T. C. A., 'Obituary,' *British Medical Journal*, 2 (August 26, 1893), p. 496.

82 J.-M. Charcot, *Lectures on Diseases of the Nervous System*, 1st series, tr. George Sigerson (New Sydenham Society, London, 1877), 326 pages.

83 J.-M. Charcot, *Lectures on the Diseases of the Nervous System*, 2nd series,

tr. and ed. George Sigerson (New Sydenham Society, London, 1881), 400 pages.

84 Jean-Martin Charcot, *Clinical Lectures on Senile and Chronic Diseases*, tr. William S. Tuke (New Sydenham Society, London, 1881); *Lectures on the Localisation of Cerebral and Spinal Diseases*, tr. and ed. Walter Baugh Hadden (New Sydenham Society, London, 1883) and *On the Treatment by Suspension of Locomotor Ataxy and Some Other Spinal Afflictions*, tr. and ed. A. de Watteville (D. Stott, London, 1889). Two translations undertaken in the United States, and again concerning earlier works, also appeared in England: Charcot, *Lectures on Bright's Disease of the Kidneys*, tr. Henry B. Millard (Baillière, London, 1879) and *Clinical Lectures on the Diseases of Old Age*, tr. Leigh H. Hunt with additional lectures by Alfred L. Loomis (Sampson and Low, London, 1882).

85 J.-M. Charcot, *Clinical Lectures on Diseases of the Nervous System Delivered at the Infirmary of La Salpêtrière* , 3rd series, tr. Thomas D. Savill (New Sydenham Society, London, 1889), 438 pages. This was also the volume of Charcot's writings that the young Sigmund Freud, who had his own reasons to be interested in male hysteria, rendered into German under the title *Neue Vorlesungen über die Krankheiten des Nervensystems insbesondere über Hysterie* (Toeplitz and Deuticke, Leipzig and Vienna, 1886).

86 Jean-Martin Charcot, *Charcot the Clinician: The Tuesday Lesson*, tr. with commentary Christopher G. Goetz (Raven Press, New York, 1987) includes a case history of hysteria, in a female patient, from the first volume of the *Leçons du mardi*. Also a reprint of Savill's translation of 1889, with a substantial historical introduction by Ruth Harris, is currently in preparation as part of the *Tavistock Classics in the History of Psychiatry*.

87 An excellent introduction to these writings may be found in Richard Hunter and Ida Macalpine, *Three Hundred Years of Psychiatry, 1535–1860: A History Presented through Selected English Texts* (Oxford University Press, London, 1963), pp. 1–564. From a large secondary literature on the subject, refer to Ilza Veith, 'On hysterical and hypochondriacal afflictions,' *Bulletin of the History of Medicine*, 30 (1956), pp. 233–40; Altschule, 'Ideas about anxiety held by eighteenth-century British medical writers,' *Roots of Modern Psychiatry*, pp. 1–13; Stanley W. Jackson, *Melancholia and Depression: From Hippocratic Times to Modern Times* (Yale University Press, New Haven, 1986), chs 6, 7, 11; and Roy Porter, *Mind-Forg'd Manacles: A History of Madness in England from the Restoration to the Regency* (Athlone Press, London, 1987), esp. ch. 2.

88 Thomas Addison, *Observations on the Disorders of Females connected with Uterine Irritations* (S. Highley, London, 1830), pp. 63–85; Thomas Laycock, *A Treatise on the Nervous Diseases of Women; Comprising an Inquiry into the Nature, Causes, and Treatment of Spinal and Hysterical Disorders*

(Longman, Orme, Brown, Green, and Longmans, London, 1840) and *An Essay on Hysteria: Being an Analysis of Its Irregular and Aggravated Forms* (Haswell, Barrington, and Haswell, Philadelphia, 1840), esp. ch. 3.

89 See, for example, William John Anderson, *Hysterical and Nervous Affections of Women* (John Churchill, London, 1853).

90 Robert Brudenell Carter, *On the Pathology and Treatment of Hysteria* (John Churchill, London, 1853).

91 For a glimpse of the clash between these models, compare the two discussions of hysteria in Reynolds's *System of Medicine*. From the neurological perspective, there is Reynolds, 'Hysteria,' 1, pp. 630–46; and for the gynaecological, Graily Hewitt, 'Changes in the Shape and Position of the Uterus' and 'Disorders of Uterine Functions,' 3, pp. 749–74, esp. pp. 771–3.

92 William S. Playfair, *The Systematic Treatment of Nerve Prostration and Hysteria* (Smith, Elder, London, 1883) and 'The nervous system in relation to gynaecology' in Thomas Clifford Allbutt and W. S. Playfair (eds), *A System of Gynaecology by Many Writers* (Macmillan, London, 1897), pp. 220–31.

93 Graily Hewitt, *The Pathology, Diagnosis, and Treatment of the Diseases of Women*, 4th edn rev. (Longmans Green, London, 1882), chs 36–9, esp. pp. 549–63. Hewitt's theory was also published in pamphlet form as *The Exciting Cause of Attacks of Hysteria and Hystero-Epilepsy* (1881).

94 Cited, ironically, in J. C. Webster, *Diseases of Women: A Textbook for Students and Practitioners* (Young J. Pentland, Edinburgh and London, 1898), p. 132.

95 As even Gowers commented, 'Some disposition to hysteria is inherent, if not in all women, at least in the vast majority': *Manual of Diseases of the Nervous System*, 2nd edn, p. 985.

96 A long and explicit statement of this last point was Harry Campbell's *Differences in the Nervous Organization of Man and Woman: Physiological and Pathological* (H. K. Lewis, London, 1891).

97 The literature in this mould is large. In addition to the writings of Hewitt and Playfair, consider Robert Lee, *A Treatise on Hysteria* (J. and A. Churchill, London, 1871); Robert Barnes, *A Clinical History of the Medical and Surgical Diseases of Women* (J. and A. Churchill, London, 1873), chapters 20, 21, 24; Heywood Smith, *Practical Gynaecology: A Handbook of the Diseases of Women* (Lindsay and Blakiston, Philadelphia, 1878), ch. 5; Johnson, *Medical Lectures*, pp. 266–85; Arthur H. N. Lewers, *A Practical Text-book of the Diseases of Women* (P. Blakiston, Philadelphia, 1888), appendix A; H. Macnaughton Jones, *Practical Manual of Diseases of Women and Uterine Therapeutics*, 4th edn. (Baillière, Tindall and Cox, London 1890), chapter 21; Robert Barnes, 'On the correlations of the sexual

functions and mental disorders of women,' *British Gynaecological Journal*, 6 (1890), pp. 390–406, with discussion on pp. 406–13; Thomas More Madden, *Clinical Gynaecology: Being a Hand-book of Diseases Peculiar to Women* (J. B. Lippincott, Philadelphia, 1893), ch. 41; and W. Balls-Headley, 'The etiology of the diseases of the female genital organs' in Allbutt and Playfair (eds), *System of Gynaecology*, pp. 112–51. For the views of alienists, consider T. S. Clouston, *Clinical Lectures on Mental Diseases*, 4th edn (Lea Brothers, Philadelphia, 1897), lectures 14–15, esp. pp. 527–30; and George H. Savage, *Insanity and Allied Neuroses: Practical and Clinical*, 7th edn (Cassell, London, 1896), ch. 5, esp. pp. 89–90. A rare critical voice from within the profession may be found in the American William Goodell's, *Lessons in Gynecology*, 3rd edn (F. A. Davis, Philadelphia and London, 1890) which described uterine overpathologization as 'the crying medical error of the day' (p. 519).

98 Hewitt, *Pathology of Diseases of Women*, p. 565.

99 The heyday of European interest in hysteria paralleled the high point of the practice of gynaecological surgery, including extirpative surgery for nervous and mental disorders. Mercifully, these practices were never as widespread in Britain as in either Germany or the United States. (As G. J. Barker-Benfield has suggested, this was perhaps because doctors in Britain remembered the public ridicule and professional censorship of Isaac Baker Brown, the 'surgeon-accoucheur' who in the early 1860s had performed clitoridectomies on hysterics, epileptics, and the mentally deficient). Among *fin de siècle* gynaecologists in Britain, the propriety of such procedures was in fact sharply debated, with many physicians deploring the practice. However, Robert Lawson Tait, the most eminent British surgeon of the day and pioneer of the ovariotomy for organic pathologies, refused to condemn the operation in the treatment of nervous ailments. Furthermore, the record of medical publications indicates clearly that a number of British gynaecological surgeons continued throughout the period to intervene surgically in recalcitrant cases. While the relationship between printed medical materials and clinical practice is difficult to ascertain, it is also likely that many more such procedures were conducted than were reported in the medical press. In at least one instance, that of Dr Francis Imlach in 1886 at the Hospital for Women in Liverpool, this led to a medical enquiry and civil court case initiated by a patient. The relevant sources include Baker Brown, *On the Curability of Certain Forms of Insanity, Epilepsy, Catalepsy, and Hysteria in Females* (Robert Hardwicke, London, 1866), chapter 3; Lawson Tait, *The Pathology and Treatment of Diseases of the Ovaries* (William Wood, New York, 1883), pp. 232–321, 326–9; Thomas Savage, 'Oöphorectomy; being a series of twenty-five consecutive successful cases,' *Birmingham Medical Review*, 4 (1881), pp. 147–61; W. Walter, 'A

case of hystero-epilepsy cured by excision of both ovaries,' *Liverpool Medico-Chirurgical Journal*, 2 (1882), pp. 380–5; Francis Imlach, 'A case of hystero-epilepsy of twenty years' duration, treated by removal of the uterine appendages,' *British Medical Journal*, 1 (April 7, 1888), p. 740; R. T. Smith, 'On hystero-epilepsy,' *British Gynaecological Journal*, 5 (1889), pp. 69–78; Draper, 'Hysterical vomiting and other cases treated by hypnotism. Oöphorectomy in two of the cases,' *British Gynaecological Journal*, 7 (1891), pp. 338–45; and Hewitt, *Pathology of Diseases of Women*, pp. 692–7. The story of *Casey v. Imlach* is related in *Lancet*, 2 (1886), pp. 470–1, 603–4, 748–49, 1147–49.

100 Addison, *Observations on Disorders of Females*, p. 95; Laycock, *Treatise on Nervous Diseases of Women*, p. 8; Lee, *Treatise on Hysteria*, p. 21; Hewitt, *Pathology of Diseases of Women*, p. 529. See John S. Bristowe, *A Treatise on the Theory and Practice of Medicine*, 7th edn (London, Smith, Elder, 1890), pp. 1138–51; and Savage, *Insanity and allied Neuroses*, p. 87.

101 Addison, *Observations on Disorders of Females*, pp. 94–5; Hewitt, *Pathology of Diseases of Women*, pp. 527–8, 547.

102 Showalter, *Female Malady*, chapter 7; Martin Stone, 'Shellshock and the Psychologists,' in Bynum, Porter, and Shepherd (eds), 2, pp. 242–71; Harold Merskey, 'Shellshock' in G. E. Berrios and H. L. Freeman (eds), *1841–1991: 150 Years of British Psychiatry* (Royal College of Psychiatrists, London, forthcoming).

PART III

Science and Feminism

8

'Life' as We Have Known It:
Feminism and the Biology of Gender

Lynda Birke

INTRODUCTION:
CONTEMPORARY FEMINISM AND WOMEN'S BIOLOGY

Contemporary feminism has generally opposed any attempts to attribute women's oppression to biology. Examples of crude biological determinism have not been hard to find, implicating everything from how we feel about our children or our sexuality, to whether we are capable of particular jobs. These arguments have said that gender divisions are natural, the product of an underlying biological imperative that angry women could do little to change. For that reason at least, feminists tended to reject such determinism: women's liberation, after all, was a social movement predicated on the idea of change. In the first flush of enthusiasm in the early 1970s, women did not intend to remain subservient to men, and they argued strongly that 'sex roles' had nothing to do with biology and much to do with social learning.

Beside the simple rejection of biological determinism, however, emerged other responses. Some feminists, for example, have argued that it is not enough just to reject the ideas themselves; we have also to tackle the institution of science itself, that generates or perpetuates those ideas. Other feminists began to question the emphasis on social construction of gender. The realities of the body, they noted with concern, had somehow vaporized, as though they had little or nothing to do with women's experience.

Feminists responded to this omission in various ways. Some seemed to return to arguing essentialist beliefs in defence of particular politics; female biology and its potential for reproduction, for example, provided for some writers the core of female nature – the 'bedrock of who I am'[1] – which could not be ignored or subsumed into social learning. Female biology is also, others argued, in danger of a potential technological takeover, as (largely male) scientists find more ways to intervene in reproduction and so further disempower women.[2] Yet others believed that women were somehow closer to nature, more intuitively attuned to it, and so better able to defend the earth and its life than men.[3] By the 1980s, feminism was caught between the need to reject crude biological determinism and the need to recognize that nature, including our material bodies, helps to shape our experience.

Yet ambivalence about women's relationship to their biology is not unique to contemporary feminism. In this chapter I shall examine some of the views expressed in earlier waves of feminism and how these relate to contemporary feminist attitudes towards biology. Although the content of criticism inevitably differs historically, as does its social effect, there are some recurring themes in feminist writing. One approach, evident in some nineteenth-century and contemporary feminist criticism, is to contest the science on its own ground, to point out the hidden assumptions and poor methodology in, say, research into gender difference. A stronger, more radical, version of this approach goes beyond critiques of 'bad science', and emphasizes the social context of science, arguing that social and political relations are embodied in both its practice and content.[4] The third strand is also critical of science and prevailing beliefs in biological determinism, but criticizes by reinterpreting the data; this theme, for example, might accept a given gender difference but interpret it by arguing for female superiority. While I am critical of such an essentialist view of gender, it has sometimes been politically expedient for feminists to make this claim.

Contemporary feminism continues to be ambivalent about women's biology. In wanting not to be defined by it, feminists initially rejected it, and argued – as did many others on the left – for social constructionism. On the whole, most feminists (particularly in Britain and the USA) tend still to do so. Yet that emphasis continues to provoke reaction, not least because it seems to deny precisely those bodily functions by which women have always been defined.[5]

This dilemma serves to underline how much feminists, like others, tend to remain trapped in the old nature/nurture dichotomy. Either we have seemed to reject 'nature' and argued for 'nurture' or we have implicitly accepted that women have a specific, essential nature. One reason for this entrapment is that 'biology' tends to be seen as fixed, as an identifiable entity; and feminists have been just as good at making this assumption as everyone else. What feminism needs now, I will argue, is to develop a less rigid view of women's (indeed, of human) biology and to see it in terms of transformation, or processes of change.[6] This is one component of what some feminist writers have described (albeit sometimes controversially) as a 'feminist science'. And it is these developments – conscious attempts to change the structure of knowledge itself – that seem particularly to characterize contemporary feminist discourse, to differentiate it from earlier waves of feminism.

DEFINING BIOLOGY

Part of the problem for feminists has been the elusiveness of some of the terms. Not only is 'biology' used to describe a set of physiological functions that characterize a particular species or individual; it is also used (sometimes interchangeably) to define a particular discipline or study – the 'life sciences', by which we attempt to describe the 'life' of 'nature'.

Biological determinism includes arguments that define social inferiority in terms of an underlying biological basis; in relation to gender differences, that may be hormonal, developmental, evolutionary or whatever. This biological basis is alleged to determine patterns of development within the individual. Becoming a woman comes from various kinds of 'whisperings within', to borrow a phrase from contemporary sociobiology.[7] Biological determinism, moreover, defends and maintains relationships of power, telling us not simply that women and men are different due to their biology, but that one is superior to the other. Thus, women should not be bank managers or presidents because their 'raging hormones' make them less rational; patriarchy is inevitable because male dominance is rooted in 'male' hormones.[8] Another problem with the deterministic literature that feminism criticizes is that it elides the concept 'woman' with that of 'female'. 'Female' has a meaning in relation to the functioning of human bodies, in that biologists understand it to mean organisms that produce eggs. 'Woman', on the other hand, has social meaning, particularly in the feminist lexicon; its meaning is socially constructed and historically contingent.

One distinction upon which much contemporary feminist theorizing rests is that between sex and gender. Sex is held to be a primary category, the division into two biological sexes, male and female. Culturally, we assign an infant to one or other at birth and treat it accordingly. Gender then follows, as a dichotomy built upon that of sex. The sex/gender distinction has presented feminist theory with problems, however, because it often seems to deny much of our biological experience; for that reason, it has been subject to increasing criticism from feminists.[9]

That distinction, moreover, is predominantly a contemporary one. Modern feminists, Donna Haraway has pointed out, construct the notion of gender as 'positioned difference', as relational.[10] Looking back at earlier feminists' reactions to biological determinism requires that we acknowledge a somewhat different use of the concept of sex from the concepts typically used by feminists now, as much of the nineteenth-century writing arguing for or against biological determinism relied on a rigid concept of sex which included what we might now call gender.[11] In 1891, for example, Eliza Lynn Linton wrote an article arguing against feminist demands, and for acceptance of women's role in maternity. Sex was the basis of her anti-feminist stance: 'This question of woman's political power is from beginning to end a question of sex, and all that depends on sex – its moral and intellectual limitations, its emotional excesses, its personal disabilities, its social conditions. It is a question of science . . . And science is dead against it.'[12] Sex, the biological substrate, was the overriding category in discussions of 'the woman question' (both feminist and antifeminist) in the late nineteenth century, as well as in emerging investigations of homosexuality.[13] The more mutable notion of gender as a product of social learning was much less evident than it is today.[14] In discussing ideas of biological determinism, however, I intend to use the modern concepts of sex and gender, despite their problems.

FEMINIST OPPOSITION TO BIOLOGICAL DETERMINISM; CRITICIZING THE SCIENCE

Many feminists in the nineteenth century objected to biologically determinist arguments about women in much the same way that feminists do now. One important strategy has always been to point to the lack of any evidence for biological determination, to appeal to the science itself. This usually takes the form of arguing that biologically determinist statements rest on 'bad science', with the implicit corollary that 'good science' is both possible and likely to lead to different interpretations.

The notion that masculinity or femininity represent properties of particular individuals, women and men, is in part a product of the nineteenth century. Prior to that, beliefs about gender may have been less rigidly dichotomous, and more subject to change. Ludmilla Jordanova, for example, has argued that writers of the eighteenth century were more inclined to believe that everyone had some elements of both masculinity and femininity; these elements were in constant struggle, but were not necessarily the property of any one individual.[15] The debates about the desirability of education for women echoed these beliefs; women may have been weak or silly, but this was the product of social injustices that restricted women's experience rather than the product of inborn female nature. The sexual double standard, similarly, was interpreted early in the eighteenth century as 'meaning that the rules were different for men and women, not that there was a natural difference between men and women;' later, it was natural differences that were held to blame.[16] But, by the end of that century, and certainly in the nineteenth century, beliefs about gender had become more rigid and dichotomous.

Coinciding with these changes, the science of biology was emerging from traditional natural histories, and the structural units of life – cells, organs and so on – became the focus of study.[17] In this context, once differences between individuals come to be seen as 'natural', it is but a small step to seek the causes of difference within those individuals, within the lower-level units which they contain. What the causes of women's alleged emotional vicissitudes are believed to be has, of course, changed according to the dominant theories of biology at any one time – from the uterus, to ovaries and thence to ovarian hormones, for example.[18]

By the middle of the nineteenth century, then, medical and biological science was making claims about the fixity of gender, and about the inherent weakness of women. Women were repeatedly told in the pages of medical journals of their predisposition to sickness, or about the dangers of expending so much energy that reproductive functions would suffer.[19] The biological arguments took roughly two forms; women were either alleged to be weak or inferior because of the peculiar features of their anatomy or physiology, or their physiology was itself inherently so weak that it was easily disrupted. At a time when women were struggling to enter higher education, a major source of such disruption was held to be education itself. Thus, Walter Johnson, writing in 1850 in a popular medical book entitled *The Morbid Emotions of Women* alleged that 'the grand cause of hysteria . . . that which melts the women of England into powerless babes . . . and shatters their intellect . . . and, by destroying reason, level them with those that chew the cud – this grand traitor and foe to humanity is Polite Education.'[20]

Not surprisingly, feminists took issue with such claims. Mrs E. B. Duffey, for example, published a rejoinder in 1874 to Edward Clarke's much-discussed *Sex in Education*. Clarke, a professor of medicine at Harvard University, had written the book, which argued the constraints of female biology, as a response to those American colleges that were opening their doors to women. Feminists, not surprisingly, argued back: Duffey, for example, claimed that: 'Instead of discovering that the physical ills of woman result from her following man's method of life and study, I have become convinced that they first originate from, and are afterward aggravated by, a course of life which recognizes an element of imagined feminine weakness and invalidism to which it is necessary to yield.'[21] Duffey's tactic was to criticize the assumptions that Clarke made about biology, to criticize the 'facts' on which his claims were made. So, for instance, when Clarke claimed that each generation, following 'the law of hereditary transmission, has become feebler than its predecessor', Duffey pointed out that actuarial tables indicate a trend towards greater longevity – hardly an indicator of enfeeblement.

Antoinette Brown Blackwell was another feminist whose opposition was based on attacking the validity of scientific claims. In 1875, she published *The Sexes Throughout Nature* – written in response to the explanations of gender difference offered by Darwin and Spencer – and in which she pointed out that 'It is to the most rigid scientific methods of investigation that we must undoubtedly look for a final and authoritative decision as to woman's legitimate nature and functions. . . . But science has not yet made the feminine constitution and its normal functions a prolonged and careful study.'[22] Blackwell argued at length against the claim that, through evolution, men had become superior, more highly developed beings. Beliefs in the complementarity of the sexes were the cornerstone of much biological determinism at the time, including ideas about the evolution of sex itself.[23] Blackwell's strategy of response, like that of many other feminists, relied on similar beliefs about gender difference to those of her opponents, but used evolution to argue a more feminist case. For Blackwell, nature was the basis not of women's inferiority, but of equality within difference.

A more sustained challenge to prevailing ideas of biological determinism came by the end of the nineteenth century, particularly from psychology, as more women were accepted into graduate schools. Women such as Helen Thompson Woolley and Leta Stetter Hollingworth in the USA carried out research on psychological differences between men and women and challenged the widespread assumptions that differences were large and rooted in biology.[24] In doing so, they forced male colleagues to reappraise at least some assumptions. As Rosalind Rosenberg has observed:

> In focusing on sex, they contributed in an important way to the trend towards environmentalism in American psychology. For most psychologists, the female sex represented the limiting case in how seriously they could question biological determinism. To doubt the physiological basis of feminine psychological differences was considerably to loosen the keystone of the arch of biological determinism.[25]

But, from the perspective of contemporary feminist critiques of science, we can see how the terms of the debate have changed. To many feminists writing today, similar arguments that counterpose 'good science' – a more equitable interpretation of facts, for instance – to 'bad science' do not go far enough. For them, science should not be portrayed as morally neutral.

Radical Critiques: the Inadequacies of Science

Contemporary feminist critics of biological determinism tend to be more critical of the assumptions underpinning science; it is not only a question of how well or badly the science is practised, but also of the values that are embedded in it. Science, in these critiques, is far from being morally neutral. In part, this approach owes its origins to the radical critiques of science that emerged during the 1960s and early 1970s, in the wake of the Vietnam War. Feminist critics today usually point to the ideological function served by biological determinism, and how such arguments arise from a science embedded in a particular society.[26]

Yet awareness of the political context is not exclusive to late twentieth-century distrust of science. Frances Power Cobbe, for example, questioned in 1888 why science should address 'the woman question' at all. 'Of all the theories current concerning women', she wrote, 'none is more curious than the theory that it is needful to make a theory about them.'[27]

Cobbe was not only an outspoken feminist: she was also an ardent supporter of the Victorian antivivisection movement. This was the cornerstone of late nineteenth-century radical criticism of science, evoked largely in response to what was perceived as the threat of the growing power of the medical profession. Science was seen as meddling dangerously in nature and health. In this perception, the antivivisectionists were allied with movements opposing vaccination and the implementation of the Contagious Diseases Acts.[28]

Feminist involvement in these forces of opposition was strong. The treatment of women by medicine that was enshrined in the Contagious Diseases Acts paralleled, in many ways, the treatment of animals within physiological research. And feminists were particularly active in both campaigns. Still, neither were exclusively feminist concerns; indeed, the antivivisection movement drew support from people who were avowedly antifeminist, and some feminists were opposed to antivivisection.[29]

Links between antivivisection and the concerns of feminism are still made today, although antivivisection as a form in which feminists particularly protest against the context of science has disappeared. The links, rather, have been forged mainly through the position of women, and animals, as objects of science; women, like animals, are closer to nature, the subject-matter of the science of biology. The links are made most clearly in ecofeminist writing, the concern of which is primarily with the common exploitation of nature/the environment and women.[30]

Although nineteenth-century antivivisectionists were often feminists, the equivalent social movements today overlap much less; there is relatively little concern with feminism in the animal-rights literature, and similarly relatively little concern in feminist circles with issues of animal rights.[31] In particular, the rapidly expanding literature on feminist science and epistemology has hardly considered the animal question at all.[32] One reason for this, Barbara Noske has suggested, is that contemporary feminism is only superficially critical of science: it takes at face value at least some of its premises about human relationships to nature. In particular, 'feminists have uncritically embraced the subject-object division between humans and animals, an attitude inherited from Western scientific tradition as a whole . . . Objectifying portrayals of animals continue to be accepted as "true".'[33]

A second effect – although not a necessary one – of the identification of women with nature/animals evident in much ecofeminist literature is, however, a tendency towards essentialism, of which feminists are rightly critical in other contexts. This, too, was a feature of at least some nineteenth-century feminist writing about women and their biology. But it was arguing more than that women were essentially different from men.

THE SUPERIORITY OF WOMEN

One response to claims that the behaviour of men and women is biologically determined is not to question their biological basis, but instead to advocate a 'feminist' version. In the nineteenth century this often took the form of essentialist arguments, particularly about what women saw as their better qualities (such as nurturance or moral rectitude). Some even extended this to an 'apocalyptic feminism' (particularly in the USA) that saw women's allegedly better qualities as giving them a leading role in the creation of a better world.[34] Eliza Farnham, for example, used physiology (and the idea of a chain of being) to argue in 1864 for female superiority: "Life is exalted in proportion to its Organic and Functional Complexity; Woman's Organism is more Complex and her totality of Function larger than those of any other being inhabiting our earth; Therefore her position in the scale of Life is the most exalted – the Sovereign one.'[35]

The notion of a special female nature, or even superiority, was particularly prevalent towards the end of the nineteenth century, emerging out of the increasing separation of public and private spheres.[36] Feminists were quick to challenge the notion of 'true womanhood' that restricted women to the domestic sphere; but they did so without ever quite losing the basis of that notion in essential natures. The belief that women, by nature, were more moral underlay, for example, the involvement of feminists in the purity campaigns of the late nineteenth century and the related campaigns against the Contagious Diseases Acts. Women and men were widely believed to have quite different sexual natures: men's sexuality was spontaneous and easily aroused, while women's was often thought to be nonexistent. For many feminists, this was the basis of the belief that women were 'naturally' more moral, and so should be custodians of morality. The resultant purity campaigns, unlike later feminist activities, did not try to open up opportunities for women to enter men's domain; rather, they were geared to encouraging men towards the kind of chastity expected of women.[37]

Feminists did, however, challenge the notion of separate spheres itself and its underlying biological determinism. Olive Banks has suggested that as feminists did so, particularly in the USA, 'The cult of domesticity became transformed into the ideal of female superiority, and the doctrine of separate spheres into the attempted invasion of the masculine world not simply by women but, potentially, even more revolutionary in its impact, by womanly values.'[38] There were two noteworthy consequences of this emphasis. In the first place, the stress laid on 'maternal mystique' led many feminists into involvement with the eugenics movement.[39] Secondly, emphasizing women's 'special qualities' implied the kind of biological determinism that feminists in other spheres were attacking.

Most feminists at the end of the nineteenth century expressed some support for eugenics, both in Britain and the USA. The eugenics movement relied on ideas of social Darwinism and the belief that the physical and mental level of humanity could be raised by encouraging those who were more 'fit' to breed (while simultaneously discouraging the less 'fit' from reproducing). The appeal for feminists was twofold; in the first place, eugenics advocated control of the population, a demand which echoed the voices of feminists involved in campaigns for birth-control (Marie Stopes, for example, developed an interest in sex and birth-control through an interest in eugenics).[40] Feminists were also drawn to eugenics because of its advocacy, on Darwinian grounds, of greater freedom of choice for women in marriage. The eugenics movement was, however, ambiguous in its support for feminism, tending to place motherhood and emancipation in opposition to one another, and opposing reforms that would allow women independence.[41]

Contemporary feminists tend to abhor any idea that smacks of eugenics; not only does 'eugenics' conjure up images of Nazi selection of the 'unfit' and the 'fit', but it implies a very rigid biological – genetic – determinism. With twentieth-century hindsight, it seems strange that so many feminists around the turn of the century did, even if only partly, support eugenic views. One reason was the glorification of motherhood that it implied. But another possible reason may – speculatively – have been that feminists at the time did not use the kind of strictly genetic framework that we now associate with eugenics. The feminist writer Charlotte Perkins Gilman (1869–1935), for example, had eugenic beliefs, but she also advocated a neo-Lamarckian (rather than Darwinian) view of human evolution. Ann Palmeri has suggested that this allowed a more environmentalist or culturally-influenced conception of human development – which, by implication, was less rigidly biologically determinist than the more familiar tenets of social Darwinism.[42] So it is possible that at least some nineteenth-century feminists did not view women's place in society in quite the deterministic fashion that the 'cult of motherhood' now implies.

Still, speculation aside, the notion that women, particularly in their role as mothers, were superior, does undoubtedly lend itself to essentialism. This tendency has bedevilled feminist thought at various times, and does so still. At a time when women were beginning to enter the professions in increasing numbers, to imply that women had an essential female nature was problematic. Yet we must recognize that to imply an essential nature was sometimes a useful strategy. Nineteenth-century feminists extolling the superiority of women sometimes did so in order to argue for better education and women's suffrage; like Mary Wollstonecraft a century before, they believed that education enabled women to be better mothers. Similarly, Radclyffe Hall followed Havelock Ellis in her belief that homosexuality was inborn, an essential quality of individuals half-way between men and women; to Hall, to encourage this belief was to encourage greater social tolerance of a beleaguered minority.[43]

It was, however, the push for equal rights in the public sphere, particularly in the professions and higher education, rather than feminist reliance on essentialist theories, that shifted emphasis away from the notion of 'separate spheres'.[44] It was also the equal-rights tradition that, by enabling women to enter scientific professions, broke the link between women and their opposition to science in the antivivisection movement.[45] Not until much later would feminists again begin to question science itself in any systematic way.

THE DILEMMA RETURNS;
CONTEMPORARY FEMINISM
AND BIOLOGICAL DETERMINISM

As contemporary feminism emerged in the 1960s, partly out of other movements of social protest, it challenged assumptions that existing gender relations were both right and natural. As feminists of previous generations had done, one response of contemporary feminists was to reject any inference that biology might be involved, and to point to the power of social structures – including science itself – in generating gender divisions.

Necessary though that response was politically, it generated reaction both from within feminism and from its opposition. The core of the problem was that the equal rights tradition, which emphasizes the social construction of gender, minimizes differences. There have been two significant effects of this on contemporary feminist thought: first, it implicitly ignores the reality of the body and bodily differences. Second, it also minimizes differences between women. During the 1980s, both these have become foci of feminist attention, sometimes generating much dissent.

Antifeminist response during the 1970s was prompt; feminists were accused of a naïve environmentalism and of being opposed to any involvement of biological processes in the development of gender.[46] Feminists themselves responded in various ways. One response was to emphasize that our understanding of our biology is itself socially constructed; if women suffer from menstrual problems, for example, this is the result of social expectations of menstruation.[47] Another response, as I noted earlier, was a return to a feminist essentialism which not only sometimes celebrated women's biological functions, but also implied that women's 'nature' was accordingly inherent.[48] Twentieth-century feminists, like their predecessors, were caught in the old impasse; if not nurture, then nature.

Contemporary feminism has perhaps gone further than its predecessors in trying to meet that challenge. Recent French feminist writing, for example, has tried to incorporate a concept of the body and its meaning into feminist accounts of gender and psychoanalysis.[49] The French feminist Luce Irigaray, for example, argues that women's oppression lies not merely in social and economic structures, but in the very language and processes of thought that epitomize our culture. To break free, women must learn to understand the psychic roots of our differentness, which, Irigaray believes, requires finding new ways of describing the female body and its sexuality. Yet much of this writing has remained firmly trapped in essentialism: as Lynne Segal has pointed out, this 'account of sexual experience in the female imaginary is one which exists outside the changing social institutions like the family . . . [and is] most readily interpreted as strengthening and celebrating traditional gender ideologies of fundamental biological difference between women and men.'[50] One factor that contributes to that entrapment is that theories of gender tend to be based on an assumption of passivity: we become victims of either cultural conditioning (including language) or the imperatives of our genes. But the fixity is most apparent in relation to how we think about our biology; most people simply do not think about 'biology' in terms of process and change. So to admit 'the body' into feminist theorizing has typically implied static constraints; the body is simply there. And just because feminism has always been about challenge and change, stasis does not fit well with our theories.

There are obviously some things about the body that are relatively fixed. Unless and until reproduction is possible entirely without human bodies,[51] then female bodies will bear children. The anatomical structures that go with that are a relatively constant feature. Yet there are ways in which our understanding of even our reproductive 'biology' is not static. The significance attached to reproductive structures, to bodily functions, may change, for example. As I have pointed out above, feminist responses to biological determinism in the nineteenth century assumed much more emphasis on the concept of 'sex' (rather than 'gender') than we would make now. While we still tend to believe that the fact of reproduction will inevitably impose some constraints on, say, any attempt to create a less gendered society,[52] our image of those constraints is much less powerful than that of our nineteenth-century predecessors. One reason, of course, why sex was the organizing concept was simply that women could then less easily escape the imperatives of the female body: the best contraception was simply to avoid marriage. Today, with arguably more freedom to make reproductive decisions, we see reproductive potential as more malleable.[53]

Transformation and change, moreover, *are* part of biological processes. 'The body' is never a static, unchanging thing, nor does it exist in splendid isolation from the rest of our lives. What we need to take greater account of is the extent to which bodily processes both affect, and are affected by, our experiences. Gender is not an edifice built upon a biological dichotomy (sex); rather, our experiences in a gendered world can themselves influence those biological processes.

Feminist writing has recently begun to include discussion of how biological processes throughout our lives might become part of how we conceptualize both sex and gender.[54] There are several potential consequences of this: first, to understand our biology as transformation can contribute to the deconstruction of the sex/gender duality that is currently part of feminist discourse.[55] Secondly, breaking down the notion of biological sex as fixity allows us to consider the uniqueness of individual developmental pathways: each of us traverses a pathway comprising both biological and sociocultural development. Simplistic dualities like 'sex' and 'gender' gloss over that variation, and ignore individual developmental histories. This shift is particularly salient in the context of contemporary emphases on differences within feminism.

Thirdly, thinking of biological processes in terms of transformation provides, in principle, a way of becoming less deterministic. That is, it could offer us a way of avoiding the pitfalls of biological determinism while simultaneously recognizing the importance of the body in our experiences as gendered beings. For women to define and theorize their own bodily experiences would be a transformation in itself – 'life' as we have known it.

Still, to conceptualize bodies and biological processes in this dynamic way is not easy. I should stress at this point that by talking about 'transformation' I do not mean something that is synonymous with 'interaction'. It is now quite fashionable among biologists to disavow extreme biological determinism and to point out how, for example, genes and environment 'interact'. They do; but I am arguing for a stronger meaning than that implied by 'interaction', which usually allows, in principle, that components can be separated (this is like saying that oil and water interact if mixed up; but they can easily be separated).

Susan Oyama has stressed that there remains in scientific thought a strong residue of preformationism,[56] such that any apparent uniformity is still seen to be provided by constraining genes. Interaction, she has pointed out, is no answer to the nature/nurture problem: 'A solution that combines encoded nature with various doses of contingent nurture is no solution at all.'[57] Ideas of transformation are certainly evident in the history of biology; eighteenth-century preformationism was routed by emerging ideas of epigenesis in embryology, and, most noticeably, evolutionary theory is fundamentally about change. But somehow notions of transformation do not always inform either other areas of biology or popular conceptions of it. Thus the behavioural sciences, on which much feminist theorizing draws, have not tended to espouse epigenetic ideas, and popular beliefs – armed with 'pop' sociobiology – now equate evolution with essences.[58]

Some version of epigenesis, of developmental transformation in which the individual and her/his body is an active agent, is essential to thinking about the body in general or about sex/gender in particular. Even sex itself, the anatomical dichotomy by which children are classified at birth, is itself something that emerges from the interaction of various processes and influences during fetal life: it is not given in the genes or in levels of fetal hormones.[59] Nor are our adult bodily experiences given: the events of my menstrual cycle, say, are as emergent from my bodily engagement with the world, as is my desire to write articles. That is not, I would emphasize, to say that my experiences of that cycle are entirely socially constructed; what I am saying is that the 'biological' events themselves are a product of that engagement.

Transformative viewpoints are, I would argue, essential if we are to understand how gender is constructed without denying the reality of the body. They are also essential to feminist interpretations of how science itself might change.[60] Feminist criticisms of science have taken various forms, although most critics emphasize two points. The first overlaps with other radical critiques of science, namely, that science as both practice and body of knowledge incorporates and epitomizes the values of the larger society. The second follows from that: if patriarchal and capitalist science is what develops in this society, then what feminists want to work for is the creation of a feminist science – a 'successor science' to what we have now. Yet in the pluralist politics of feminism in the 1980s, creating 'successor science' projects has been an idea fraught with problems. As Sandra Harding has pointed out, many of these projects are universalizing; the experience of all women becomes the basis for a feminist 'successor science'.[61]

Much of the discussion about women, feminism and science has argued the point that women, because they share a non-privileged position in the world, will inevitably bring to science a different perspective, a different way of knowing. So, the argument has gone, what we should seek is a science that incorporates that way of knowing as well as the dominant viewpoint.[62] But, Harding argues, 'there is no "*woman*" to whose social experience [these viewpoints] can appeal; there are, instead "*women*" of all kinds of different backgrounds and experience.'[63] Recognizing these 'fractured identities' is crucial both to further development of feminist theory and to our theorizing about science.

Universalizing, moreover, is just what science does in its production of knowledge (including that about women). Thinking about the body in terms of individual developmental trajectories has the merit not only of allowing for diverse endpoints – fractured identities – but also of challenging the wisdom of conventional explanations in science. By assuming universality, science rarely admits to differences – such as those between individuals, for example.[64]

Whatever disagreements there are within contemporary feminism, there are two points on which feminist critics of science would broadly agree: first, there is a need to think about ways of transcending the mire of nature versus nurture and the associated sex/gender duality that has determined so much of our theorizing. And, secondly, science itself must change; it is a fiction to think that science speaks to a universal experience of 'man'.

It is, moreover, these two challenges that take contemporary feminism rather further than its predecessors. Biological determinism is still a problem that feminists have to tackle wherever and whenever it occurs; and there is still a danger of essentialism even within feminist thought. Where contemporary feminism goes noticeably further is in its insistence that science is inherently problematic. Some of our nineteenth-century predecessors challenged the authority of science in the struggle for antivivisection. But they did so partly on rather essentialist grounds: because of their nature, women were more moral and not driven to such cruelty. Indeed, it was that belief that helped the demise of the movement; as more women fought to enter medicine on equal terms with men, more of them came to defend the use of animals in experiments.[65]

The challenge from the antivivisectionists at the turn of the century was, moreover, one addressed primarily to the practice of science; it raised fewer questions about the content and ideology of science. A corollary of that is that if the way in which men practised science in the laboratories could be changed, then we would have a 'better' science. But that does not go far enough for many feminist critics today. And perhaps that is the strength of contemporary feminist criticisms: their ultimate goal is to transform, to change nothing less than the power of science itself – its assumptions, practices and relationship to the world. That is no small task. But asking the impossible is what feminism has always been about.

NOTES

1 This emphasis forms part of what some writers, particularly in the USA, see as a return to conservative values. See Judith Stacey, 'The New Conservative

Feminism', *Feminist Studies*, 9 (1983), pp. 559–83. Stacey cites as an example Jean Bethke Elshtain's *Public Man, Private Woman* (Princeton University Press, Princeton, 1981).

2 See Gena Corea, *The Mother Machine* (Harper and Row, London, 1985).

3 See, for example, Leonie Caldecott and Stephanie Leland (eds), *Reclaim the Earth* (The Women's Press, London, 1983).

4 This critique forms part of a wider radical science criticism, embodied in Britain in organizations such as the British Society for Social Responsibility in Science, and in publications such as *Science for People* and the former *Radical Science Journal*.

5 For further discussion see Lynda Birke, *Women, Feminism and Biology: the Feminist Challenge* (Wheatsheaf, Brighton, 1986).

6 While a few radical biologists may endeavour to develop ideas of transformation and process in biology, the vision of the biological body more commonly found in biological theorizing is a fixed one. And that is the vision largely adopted within the social sciences. For discussion of radical ideas in biology see, for example, the Dialectics of Biology Group's *Beyond Biological Determinism* and *Towards a Liberatory Biology* (Allison and Busby, London, 1982).

7 David Barash, *Sociobiology: The Whisperings Within* (Souvenir Press, London, 1979). Barash uses evolutionary arguments – the notion, for example, that human behaviour is limited or determined by the natural selection of particular genes through millennia of evolution. Other determinist arguments may use a directly causal approach (e.g. how hormones might be said to cause aggression in adult men), or a developmental approach (e.g. how the presence of certain hormones in embryos affects subsequent brain development).

8 See Steven Goldberg, *The Inevitability of Patriarchy* (Temple Smith, London, 1974).

9 See, for instance, Lynda Birke and Gail Vines, 'Beyond nature versus nurture: process and biology in the development of gender', *Women's Studies International Forum*, 10 (1987), pp. 555–70, and Val Plumwood, 'Do we need a sex/gender distinction?' *Radical Philosophy* (1989), pp. 2–11.

10 Donna Haraway, 'Situated knowledges: the science question in feminism and the privilege of partial perspective', *Feminist Studies*, 14 (1988), pp. 575–99.

11 The significance of sex as a primary division is evident, for example, in Antoinette Brown Blackwell's 'Sex and evolution' (1875), reprinted in Alice S. Rossi (ed.), *The Feminist Papers* (Bantam, New York, 1974).

12 Eliza Lynn Linton, 'The wild women as politicians', *The Nineteenth Century*, 30 (1891), pp. 80–2; cited in Pat Jalland and John Hooper (eds), *Women from Birth to Death* (Harvester, Brighton, 1986), p. 26.

13 In the late nineteenth century there were various versions of congenital

theories of homosexuality, based on categorizations of sex. For a discussion of the role they played, see Jeffrey Weeks, *Sex, Politics and Society* (Longman, London, 1981).

14 In part, this shift from sex to gender as an overriding category was due to the impact of behaviourist social learning theories in the twentieth century; in part, too, it was due to the impact of feminism and women's greater access to work previously defined as male. See Rosalind Rosenberg, *Beyond Separate Spheres: Intellectual Roots of Modern Feminism* (Yale University Press, New Haven, 1982).

15 Ludmilla Jordanova, 'Natural Facts: a historical perspective on science and sexuality' in Carol MacCormack and Marilyn Strathern (eds), *Nature, Culture and Gender* (Cambridge University Press, Cambridge, 1980).

16 Alice Browne, *The Eighteenth Century Feminist Mind* (Harvester, Brighton, 1987), pp. 115–16, 146–7.

17 William Coleman, *Biology in the Nineteenth Century*, (Cambridge University Press, Cambridge, 1977), p. 15.

18 The discovery of hormones secreted by the ovary in the early twentieth century was, of course, pivotal to this change in biologically determinist argument. See Meriley Borrell, 'Organotherapy, British physiology and the discovery of the internal secretions', *Journal of the History of Biology*, 9 (1976), pp. 235–68.

19 The best-known example of this is Edward Clarke's *Sex in Education* (Boston, 1873). See discussion in, for example, Janet Sayers, *Biological Politics* (Tavistock, London, 1982); Elizabeth Helsinger, Robin Sheets and William Veeder, *The Woman Question: II: Social Issues 1837–1883* (Manchester University Press, Manchester, 1983); and Susan Sleeth Mosedale, 'Science corrupted: Victorian biologists consider "The Woman question"', *Journal of the History of Biology*, 11 (1978), pp. 1–55.

20 Walter Johnson, 'The morbid emotions of women', (1850): quoted in Jalland and Hooper (eds), *Women from Birth to Death*, p. 99.

21 Quoted in Helsinger, Sheets and Veeder, *The Woman Question*, p. 85.

22 Ibid., p. 104.

23 See, for instance, Patrick Geddes and J. Arthur Thompson, *The Evolution of Sex* (1889; Walter Scott, London, rev. edn, 1914).

24 See Stephanie A. Shields, 'The variability hypothesis: the history of a biological model of sex differences in intelligence' in Sandra Harding and Jean F. O'Barr (eds), *Sex and Scientific Inquiry* (Chicago University Press, Chicago, 1987), and Rosenberg, *Beyond Separate Spheres*.

25 Rosenberg, *Beyond Separate Spheres*, pp. 107–8.

26 See, for example, Brighton Women and Science Group (ed.), *Alice Through the Miscroscope* (Virago, London, 1980); Birke, *Women, Feminism and Biology*; Ruth Bleier, *Science and Gender* (Pergamon, Oxford, 1984); Anne

Fausto-Sterling, *Myths of Gender* (Basic Books, New York, 1985), and Sandra Harding, *The Science Question in Feminism* (Open University Press, Milton Keynes, 1986).

27 Frances Power Cobbe, 'The duties of women': quoted in Carol Bauer and Lawrence Ritt (eds), *Free and Ennobled* (Pergamon, Oxford, 1979), p. 66.

28 Mary Ann Elston, 'Women and antivivisection in Victorian England, 1870–1900', in N. Rupke, ed., *Vivisection in Historical Perspective* (Croom Helm, London, 1987).

29 Ibid.

30 See, for example, Caldecott and Leland, *Reclaim the Earth*.

31 For a discussion of feminism and the question of animals, see Barbara Noske, *Humans and Other Animals* (Pluto Press, London, 1989).

32 Birke, *Women, Feminism and Biology*, ch. 8, raises the issue of how a feminist science might deal with the animal issue. See also Andree Collard with Joyce Contrucci, *Rape of the Wild* (The Women's Press, London, 1988).

33 Noske, *Humans and Other Animals*, p. 114.

34 Helsinger, Sheets and Veeder, *The Woman Question*, p. xv.

35 Quoted ibid., p. 76.

36 Olive Banks, *Faces of Feminism* (Martin Robertson, Oxford, 1981), pp. 85 ff.

37 Ibid., pp. 63–84.

38 Ibid., p. 90.

39 Linda Gordon, *Woman's Body, Woman's right: A Social History of Birth Control in America* (Penguin, Harmondsworth, 1977).

40 See Greta Jones, *Social Darwinism and English Thought* (Harvester, Brighton, 1980), pp. 110–11, and Daniel J. Kevles, *In the Name of Eugenics* (Pelican, London, 1985), pp. 89–90.

41 Banks, *Faces of Feminism*, p. 99. Indeed, some people even argued, on eugenic grounds, that the best mothers would be feminists – i.e. women who are 'favoured . . . in physique and temperament' – C. W. Saleeby, 'Woman and Womanhood', (1912): quoted in Jalland and Hooper, *Women from Birth to Death*, p. 29.

42 Ann Palmeri, 'Charlotte Perkins Gilman: forerunner of a feminist social science', in Sandra Harding and Merill B. Hintikka, (eds), *Discovering Reality* (Reidel, Dordrecht, 1983).

43 See Michael Baker, *Our Three Selves: A Life of Radclyffe Hall* (Hamish Hamilton, London, 1985).

44 See Rosenberg, *Beyond Separate Spheres*.

45 Elston, 'Women and antivivisection'.

46 One of the many examples of such claims was Glenn Wilson's 'The sociobiology of sex differences', *Bulletin of the British Psychological Society*, 32

(1979), pp. 350–3. For other examples, see Birke, *Women, Feminism and Biology*, pp. 36–41.

47 See discussion in Sayers, *Biological Politics*, pp. 110–24.

48 The historical association between 'women' and 'nature' has been outlined by Carolyn Merchant, *The Death of Nature* (Wildwood, London, 1982); the association in contemporary feminist writing is perhaps most evident in ecofeminism: see Caldecott and Leland, *Reclaim the Earth*, and Collard and Contrucci, *Rape of the Wild*.

49 See, for example, Luce Irigiray, 'When our lips speak together', *Signs*, 6 (1980), pp. 66–79.

50 Lynne Segal, *Is the Future Female?* (Virago, London, 1987), p. 133. Janet Sayers makes a similar point in *Sexual Contradictions: Psychology, Psychoanalysis and Feminism* (Tavistock, London, 1986), pp. 42–6.

51 Although ectogenesis – the production of babies entirely outside the body – has provoked fears in many feminists that women will (biologically at least) become redundant, I am not convinced that ectogenesis will ever become widespread: women, after all, are cheaper. For further discussion see Lynda Birke, Susan Himmelweit and Gail Vines *Tomorrow's Child: Reproductive Technologies in the 1990s* (Virago, London, 1990).

52 See, for example, Plumwood, 'Do we need a sex/gender distinction?'

53 Indeed, that greater freedom creates what Mary O'Brien has called a 'second revolution' in consciousness around reproduction (the first being men's realization of their role): O'Brien, *The Politics of Reproduction* (Routledge and Kegan Paul, London, 1981).

54 See Birke and Vines, 'Beyond nature/nurture'. See also Lynda Birke and Gail Vines, 'A sporting chance? the anatomy of destiny', *Women's Studies International Forum*, 10 (1987), pp. 337–48.

55 For example, see Moira Gatens, 'Towards a feminist philosophy of the body' in Barbara Caine, E. A. Grosz, and Marie de Lepervanche (eds), *Crossing Boundaries: Feminisms and the Critique of Knowledges* (Allen and Unwin, London, 1988).

56 Susan Oyama, *The Ontogeny of Information: Developmental Systems and Evolution* (Cambridge University Press, Cambridge, 1985).

57 Susan Oyama, 'Essentialism, women and war: protesting too much, protesting too little' in C. Plamenbaum, A. Hunter and S. Sunday (eds), *Genes and Gender VI* (Gordian Press, New York, 1987).

58 Evolution, to many people, seems now to imply fixity – we have 'genes for conformity' or 'genes for xenophobia', according to popularized versions of sociobiology. But evolutionary theory is about *change* in populations, not fixed essences in individuals.

59 For a critique of the view that at least some aspects of morphological sex (e.g. genital structure) and gender behaviour are determined uniquely by prenatal

hormones, see Lynda Birke, 'How do gender differences in behaviour develop? A reanalysis of the role of early experience' in P. P. G. Bateson and P. H. Klopfer (eds), *Perspectives in Ethology*, vol. 8 (Plenum, London, 1989).

60 See, for example, Birke, *Women, Feminism and Biology*; Bleier, *Science and Gender*, and Harding, *The Science Question in Feminism*. This is not to say, however, that transformative views uniquely *define* the position of feminist science: on the contrary, they would form part of many radical interpretations of what science should become.

61 Harding, *The Science Question in Feminism*.

62 This is what Harding refers to as 'standpoint epistemologies' – the work, for example, of Nancy Hartsock, 'The feminist standpoint: developing the ground for a specifically feminist historical materialism' in Harding and Hintikka, (eds), *Discovering Reality*, pp. 311–24.

63 Harding, *The Science Question in Feminism*, p. 192.

64 Ibid., pp. 243–51.

65 Elston, 'Women and antivivisection'.

9

Unfathering the Thinkable: Gender, Science and Pacifism in the 1930s

Gill Hudson

The image of science has always been double-faced. On the one hand it can be seen as benevolent and as a progressive area in which people most benefit humanity. On the other, it can be depicted as militaristic and detrimental to humanity. For most scientists their work is pure and value free; for some this means that it is above and outside of conflict and thus has no association with war. For others, although they see their work as neutral and pure, they nevertheless believe that it is through the scientific enterprise that the world will advance to an eventual state where war is obsolete. Thus their own work is an opposition to war. There have also been periods of time, specifically before wars, when scientific knowledge has been generated exclusively for the use of the military.[1]

The association between war and science is extremely complex: if the issue becomes one of uses and abuses of scientific work, then the scientist is enabled to eschew any responsibility, and it is politicians and the military who are blamed for the use of scientific knowledge and products. If, however, we look at the way in which knowledge is actually generated within the scientific community we find that the clues towards understanding the relationship between science and war lie in the enterprise itself. We find an uneasy, precarious union between competition and co-operation: scientific knowledge is seen as a product of co-operation between scientists of different institutions and nations. At the same time, scientific 'secrets' are jealously guarded so that specific 'owner' institutions and nations can receive the glory associated with the discovery of new knowledge. Individual scientists are reluctant to share the work which perhaps in time will give them personal status and financial rewards. This practice of secrecy and appetite for competition is in direct opposition to the popular conception of science as a universal, communal enterprise; not only does it oppose, but, in contemporary science, it predominates. The competitive spirit, viewed by most as necessary and advantageous, has become an innate characteristic of science. It perpetuates national differences and rivalry and leads to science's deployment in war being seen as a 'natural' consequence. The use of science in war becomes less an issue of what politicians and militarists do with scientific knowledge, and more of a natural outcome which often the scientist himself precipitates and encourages.[2]

Since the Second World War, criticism has been aimed not only at those sciences connected with the military, particularly nuclear physics, but also at the use of biological and chemical systems as weapons. Scientists, medical and lay people have formed associations to promote an increase in public awareness and to lobby politicians with respect to the detrimental side of scientific research.[3] This movement has a precedent in the post First World War era, which was related to a socialist ideal of an international 'science for the people'. This ideal was a response partly to the use of science, especially chemical warfare, in the First War, and partly due to an increasingly pessimistic outlook towards capitalism fuelled by the economic crisis in the 1920s. The scientists, and the academic middle classes involved, fostered opinions which were incredibly idealistic, optimistic, and in many ways naïve. They thought that through the increase of scientific knowledge and its benevolent use, the ills of the world would be righted and war would become an anachronistic method of solving international conflict. It seemed possible that through solving scientific problems, the sick would be healed, the hungry fed and a lasting peace procured. According to Professor Delisle Burns:

> All knowledge points towards the improvement of common folk. I desire a world in which there is electric light and many other modern things and I cannot have that world if the people of China are starving and the people of Manchester are out of work. I desire a state of public service, so that by degrees the state as a killing machine and the state as a public service will come into conflict and gradually the former will disappear. You will not have to abolish war, for war will have become obsolete.[4]

The extent of pacifist activity in this inter-war period has been underestimated by historians: for most, the time was an abnormal

interlude between states of war. The impetus behind research has been
directed towards a concept of preparation for a new war following on the
disastrous terms laid down by the Treaty of Versailles. Yet closer
investigation shows that many people believed that the 'Great War' really
had been a 'war to end all wars'. However difficult international relations
became, the human suffering caused by war was too high a price to pay
for peace, and alternative methods must be found and worked for.
Research also shows that women played a very prominent part in pacifist
movements, not only in local groups, but as an international force for
peace, lobbying the United Nations and also individual leaders and
politicians.

In 1932 the Cambridge Scientists Anti-War group was inaugurated;
this group included a comparatively large percentage of women, among
them Dorothy Hodgkin, Dorothy Needham and Marjory Stephenson.
The aim of this paper is to examine the relationship between gender,
science and pacifism, using Dorothy Hodgkin's life and work as a case
study. I aim to discover whether women who were both scientists and
pacifists incorporated different methods of working in their science and
whether they used different research material as compared to their male
counterparts, who displayed a very vocal attitude towards their political
beliefs. At face value, the women showed a less overt interest in politics,
yet closer examination shows the profound sense in which socialist and
pacifist ideals were incorporated into their scientific work. First I will
briefly discuss contemporary literature on science and gender to raise the
question: is it possible to practise science in a specifically female way? (A
precedent for this type of analysis has been set. For example, Brian
Easlea, in *Fathering the Unthinkable*,[5] describes the Manhattan Project in
gender terms. He suggests that a female-oriented science would be a
benevolent and pacifist enterprise.) Then I will describe the women's
peace movement during the inter-war period, with particular emphasis on
the Women's International League for Peace and Freedom, to give the
pacifist background. I will then use the case study of Dorothy Hodgkin to
argue that her science was indeed a pacifist science, and to challenge the
notion that scientific knowledge is produced solely in the laboratory, by
focusing on the extent to which her work was influenced by her
involvement with her own family as a woman and mother.

UNFATHERING THE THEORY:
IS A FEMINIST SCIENCE POSSIBLE?

Rather than trying to reduce others to silence by claiming that what they say is worthless, I have tried to define this blank space from which I speak, and which is slowly taking shape in a discourse that I still feel to be so precarious and so unsure.[6]

So with the discourse of feminist thought; not only is it precarious and unsure, it also turns back upon itself, repeats and contradicts itself into apparent confusion and instability. There is no unified feminist theory any more than there is one unified experience or knowledge which is acknowledged by all women. However, should we be looking for one rational theory to explain a world which is irrational, confused, and which defies conventional intellectual analysis?

Contemporary feminist writers concerned with gender and science are engaged in conflict over the possibility of a distinctly female practice of science. This conflict appears to be making waves of enormous proportions because of its implications, first for working women scientists, and second, for feminist historians and philosophers of science. Most women who are working scientists dismiss the possibility, because in their eyes if their science is gendered and different, then they are not 'good' scientists compared to their male colleagues. They regard their science as value free; if their personal or emotional lives as women are reflected in their work, then that work does not constitute 'good science'. For those women working in the field of history, sociology or philosophy of science, the possibility of a female science presents problems of a different kind. They can hold to the view that gender makes no difference, in which case they then have to contend with a growing amount of literature which depicts science and its methods as male. Therefore, for them, for women to become accepted scientists they would have to adopt masculine values in the way they work and in their choice of research topics.[7] On the other hand, if it is believed that gender does alter scientific practice, then explanations of how this operates as a process have to be offered. The main problem with these explanations is that they can all be construed as being essentialist arguments: that is, that women practise science in a particular way because they are biologically 'programmed' to behave differently from men. This is problematic for two reasons: the first is that, tactically, biological determinism is a concept feminists have been consistently working to refute since the late 1960s; though, as Lynda Birke has shown, this conflict of interests has a nineteenth-century history.[8] Secondly, there is no real evidence that biological difference is the basis of anyone's behaviour.

It appears that whenever the question of gender difference is raised the criticism of essentialism is levelled at the concept, taking very little account of the actual context of the dialogue. Essentialism describes an innate difference in gender behaviour which is not the result of any social or historical experience; it describes immutable characteristics which all women at all times and in all cultures are assumed to possess. Foremost among feminists who reject this approach is Hilary Rose, who stresses women's social experience as the rationale behind explaining gender difference.[9] Her explanation is based on accounting for women's labour in the domestic sphere, which involves the reproduction of human beings and is a labour of love. Rose observes that it is an integration of manual, mental and emotional labour that women can bring to their scientific work. In her view, women's presence in science could bring about an alternative, more humanistic practice.

The ways in which any feminist analysis affects scientific practice depends on our perception of the enterprise itself. Here there are two alternatives: the first is that science is neutral and value free, and the second that science is a historical and social product.

Within the first alternative, that is, that science is neutral and value free, there are two ways of depicting science with reference to women. The first is that although women have problems with respect to education, promotion and winning acceptance as serious scientists, in the long term all that is required is that conditions with respect to these factors should improve. The second is that though science is neutral, it is nevertheless essentially masculine and detrimental towards people, especially women. The obvious outcome of this latter concept is the prescriptive argument that women should not work in the field at all. The stalemate which would eventuate would thus mean that nothing would change. In science, no less than in any other powerful and institutionalized social enterprise, the masculine is commonly objectified. Science is male, but because that male science is deemed pure and universal – because it involves fundamental truth – then feminists [and others] cannot effect alternative views.

An alternative perception of science involves depicting it as a social and historical product which people, predominantly men, have constructed.

If we view the scientific enterprise as a cultural and social activity, then we can recognize that our present practice is not the only way science could have evolved. Therefore we can envision the possibility of a different practice being carried out by a specific cultural group, and also conceptualize the possibility of future change.

Drawing on some concepts from the sociology of science, for example those offered by Harry Collins,[10] we can suggest that people's 'taken-for-granted reality' differs not only according to culture but also gender, and that men and women have different ways of perceiving the world. For example, a male scientist looking at the natural world depicts nature as object, himself as subject; in Baconian terms, nature is to be penetrated, taken apart and controlled. How can a woman scientist possibly view this concept in a similar fashion, when nature has long been personified as 'woman' in scientific and literary metaphor, and she herself is therefore part of the object which she is studying?

So if we believe that women are capable of bringing a different dimension to science, why is this so important? As previously indicated, science is not solely humane and benevolent; it also has the ability to be the most destructive force known to humankind. It is potentially profoundly damaging to both people and the environment. So the question becomes: how does gender affect whether science is destructive or benevolent? As we have seen, science is a very competitive business; it is often aggressive, and can be seen as reinforcing current forms of masculinity. As the biologist Richard Lewontin puts it, science is 'a form of competitive and aggressive activity, a contest of man against man that provides knowledge as a side product. That side product is its only advantage over football.'[11] Speculations about the possibility of a female science are more workable once we shift away from issues of essentialism and move towards an understanding of the way gender interests inform the uses of science.

PACIFISM AND WOMEN

Peace is an active quality, peace is not a negative thing, peace is not the mere denial of war. Peace is the readiness to use your brains and your goodwill to solve every problem as it arises. You have to make peace every hour of the day and every day of the year.[12]

The mothers of men are God's pre-ordained champions for the preservation of their sons from premature and violent deaths.[13]

The above quotations embody the direction, aims and philosophy of women's peace movements from 1914–40. Pacifism was not viewed solely as the absence of war; it was a practical, positive concept to be worked for, and it was specifically women's duty to bring a pacifist conscience to the attention of the world because of their domestic, nurturing role within it.

The most influential and well known group in the women's peace movement was the Women's International League for Peace and Freedom [WILPF], originally formed in 1915 as the International Congress of Women. Their first meeting, on 28 April at The Hague, was the first recorded time that women of different nations had met during wartime to express opposition to war and to discuss ways of ending military conflict. Intergovernmental peace conferences had been held previously at The Hague; one convened by the Russian Tsar in 1888, and one in 1907 supported by Theodore Roosevelt;[14] but the 1915 meeting was the first to be organized and attended solely by women. The conference was organized by the International Women's Suffrage Alliance [IWSA], of which the National Union of Women's Suffrage Societies [NUWSS][15] was the British wing. The IWSA was a well-established organization with a strong pacifist bias. In September 1914, Rosika Schwimmer,[16] a leading Hungarian feminist, then living in London, had toured the USA with the British suffragist Emmeline Pethick-Lawrence. They had presented a formulation for a mediation conference of neutral nations to President Wilson and had formed a Women's Peace Party with American feminists Carrie Chapman Catt and Jane Addams. A meeting in Amsterdam in February 1915, called by Dr Aletta Jacobs, suggested a congress of women to be held at the Hague in April, which would cut across national enmities and formulate plans for mediation. Although many women were prevented from attending the congress because of the war, there were delegates from 12 countries and representatives of over 150 organizations.[17] The resolutions from the congress concerned ways of procuring immediate peace, principles by which a permanent peace could be ensured, international co-operation and the education of children. It was decided to send envoys to rulers of both the belligerent and neutral countries of Europe and the USA, and 14 countries were subsequently visited by two groups of delegates in May and June. It must be noted here that for unattended small groups of women to be travelling through wartime Europe was very courageous and unconventional. According to the women themselves, they were well received by statesmen, but their conclusions can be judged to be over-optimistic since the world's press and public opinion frequently depicted the women's mission as either deplorable or laughable. As the war continued it became obvious that their efforts were not taken seriously enough, but the women did achieve one aim. This was to bring to the attention of governments and the public that mediation was a possible alternative to war – at least the alternatives had been discussed, but there is no doubt that the women pacifists became very disillusioned by the end of the war.

Meanwhile, in November 1915, 12 international committees of the WILPF had been set up, and a monthly journal inaugurated. For the duration of the war, contact between groups was maintained, but with great difficulty, and in May 1919, the second congress was held in Zurich, at which 16 countries were represented. British, German and American women attended. The German delegates discussed the poverty and malnutrition brought about by the continuing allied food blockade, and the first act of the congress was to telegraph President Wilson to demand that the blockade be raised. The congress discussed the Treaty of Versailles, which they considered had violated all the principles on which a just and lasting peace could be secured. However, they viewed the proposed League of Nations with approval, and consequently attended its conferences and established good relations with many delegates and officials. The inter-war period saw the growth and strengthening of the WILPF, and the inauguration of the journal *Pax International* in 1925. Delegates attended League of Nations conferences and international peace conferences to discuss disarmament and the use of science in war, and themselves organized joint conferences including other pacifist and feminist groups.[18]

THE AIMS AND PHILOSOPHY OF THE WILPF

The immediate objective of WILPF was to stop the First World War and the long term aim was to abolish the war system and replace it with an ordered co-operation among nations. Their work was to bring about 'the greatest revolution in the history of mankind', which was that conflict would be resolved by mediation not by warfare.[19] That women were responsible for this work was based on women's emotional, psychological and manual labour in the domestic sphere. The 'new world' was depicted as one in which this women's labour would override man's apparent need for killing.

As Jane Addams said of a new kind of 'heroism' which should be acclaimed: 'Not the heroism connected with warfare and destruction but that which pertains to labour and the nourishing of human life.'[20] Peace was seen as traditionally women's business:

> All these millions of wrecked men who were broken or killed upon the field of battle it was not organisations, but women, in their widely separated spheres of life, who gave them birth in suffering at a greater cost of actual bloodshed and anguish than was ever paid upon the battlefields. It is Women, Individual Woman, who pay the cost on all human life. They are individually responsible for those lives, and by their ignorance and resulting indifference they were individually responsible for the war. Women make men, men make governments and governments make war. But it all begins with the women, and comes back upon them.[21]

In the interests of motherhood, the race, the human family as a whole, women were responsible for awakening a pacifist awareness in other women, and for inculcating ideals of peace into their children's minds. They were also to be responsible for pressurizing for reforms in schools concerning the teaching of history and scripture. At a WILPF conference on the Teaching of History and Scripture in 1917, Maude Royden[22] introduced the proceedings by stating that the future of the world depended on the spirit in which the young would face its problems. Of history and scripture she complained of teaching so biased that every country claimed to be right and superior to all others. She continued that history was taught according to the abnormal and not the normal; resulting in a history of wars, great heroes, criminals and monarchs. Of life and thought, experience and habits, comparatively little was written or taught. Mr P. Gooch, a historian, called for history to be taught internationally and impartially: 'Is he [the teacher of history] to feed its [the rising generation's] mind and heart with pride and hatred, or is he to plant and foster the sense of human solidarity which enshrines the only abiding hope for the race?'[23] Of scripture it was argued that there was a need for more discernment with regard to biblical interpretation and an end to justifying war by reference to biblical conflicts.

The WILPF were also very concerned about military training in schools and conscription which, as they pointed out, was compulsory even in some neutral countries like Switzerland. It was thought in the 1920s that because England did not have conscription, it was possible to work for peace and disarmament. 'Never will we have peace in the world until military training is abolished. Our men and boys must not be taught how to kill each other but instead how to co-operate with one another'.[24] Along with disapproval of military training in schools, they were also concerned about the effect of war toys and films on young minds. Women were exhorted to show their sons that war was too terrible and sinister a tragedy to be played at as a game. War films were described as spiritual preparation for new war. A canvass of schoolchildren in 1928 on attitudes to these films showed that most boys found them exciting and felt that war gave youth the opportunity to test itself;[25] in other words it made men of the boys. All girls questioned commented on horror and sadness. This result was interpreted as being due to women not fulfilling their duty by instilling pacifism into their sons' minds. In an effort to encourage pacifism and internationalism, summer schools and camps were held by the WILPF which included adults and children.

On the political front many women in the peace movement had access to politicians in high positions by virtue of class, background or marriage.[26] Many of the leaders of WILPF were politically aware, and were impressive speech-makers, with a confidence born of moving in upper middle-class circles. Without this the movement would have achieved little influence or backing. The pacifist movement as a whole obtained more support from Labour politicians, but there were exceptions to this. For example, a Tory, Lord Cecil, before the general election of 1929, exhorted voters to vote for whichever candidate had the best record as a peacemaker. Members of the WILPF warned against the assumption that all Labour and socialist supporters were necessarily also pacifists. As Anna Kethly, a Hungarian MP pointed out, many left-wing people saw soldiers as enemies solely because they were fighting in the interests of those in power and against the working class.[27] Thus it was necessary to turn all political power against militarism and armament and to establish social and economic conditions which would make war unnecessary.

With respect to the use of science in war, all pacifists were very much aware of the dangers posed by new scientific knowledge. In September 1931 the third War Resisters' International Congress was held in Lyons, over which Fenner-Brockway presided. The most important event of the congress was the arrival of a letter from Einstein, calling for fellow scientists to refuse to co-operate in research for war purposes. The following year the seventh International WILPF Conference was held, the main agenda concerning scientific warfare. Appeals sent to all governments warned of the probability of a 'war of extermination' because of recent developments in science, including improved chemical weapons and aerial bombardment.

Despite this increasingly pervasive attitude that science was, in the main, destructive, again the double-face of science can be discerned. This time the ambivalence concerns gender. *Pax International* in 1929 carried an article by Marcell Capy on Marie Curie. Curie appears as the 'Florence Nightingale' of science. It seems that because she was a woman her science was necessarily labelled beneficial. As Capy concludes: 'And everywhere Marie Curie watches and works, that the precious secrets she has snatched from nature may serve to heal the sick and promote the progress of civilisation.'[28] Curie, as woman, is watching to ensure that her work is not used [by man?] for destruction.

It is very clear from the women's pacifist literature that women were very sceptical of trusting men to lead the way to peace. For them, peace was the only future, and that future was entrusted to the women of the world in the shape of children. The practice of creating peace was linked with the practice of creating children, and women were the sole people who could accomplish this creation.[29] As Mrs Corbett Ashby, President of the International Women's Suffrage Alliance, commented:

> For after all, if you consider the life of the normal woman, is it not a constant struggle to adjust conflicting aims in her house and family: claims of the weak and strong: of the ill and the well: of the old and the young. In our own homes we find all these conflicting claims, each one legitimate in itself, which we have to adjust for the good of the home. If we approach the problems of peace in the sense of feeling that we must adjust each nation's claim in a proper relation with the equally just claim of another country, we shall envisage peace in a practical manner.[30]

UNFATHERING THE SCIENCE – THE PRAXIS: A CASE STUDY OF DOROTHY HODGKIN

Biographies of Dorothy have so far presented the scientist/mother/pacifist as distinct entities. For example, Max Perutz described her as not only a good scientist, but also an excellent wife and mother.[31] However, because her scientific work and her mothering are seen as distinct entities, no possibility is acknowledged of a connection between her personal and public lives. Caroline Moorehead describes her as the scientist and the pacifist, again the science and the pacifism remain separate.[32] Further complicating an integrated approach is Dorothy's own view of herself in that she does not connect the personal, the political and the scientific.[33] For her, the research she carried out was by chance and curiosity, not choice, and her attitudes towards science were divorced from her personal and political life. However, as I have previously discussed, for women scientists to acknowledge personal influences on their work would be to question their work's authenticity. So when discussing Dorothy's life my aim is to show the extent of connections which did bridge the personal and public without understating the importance and worth of her scientific work.

Dorothy came from an affluent and intellectual background; both her parents were professional people, though not scientists. Her mother was a socialist and a dedicated pacifist, the latter as a result of the fact that four of Dorothy's uncles had been killed in the First War. Dorothy's interest in chemistry began early, and she was allowed to join boys' science classes at school – no doubt explaining why in later life she recognized few problems with working in a very male environment. She displayed a continued interest in Labour party politics throughout school and later at Somerville College, Oxford. After graduation she moved to Cambridge to study X-ray crystallography under J. D. Bernal.

The Crossroads: Bernal and Cambridge Science

Dorothy arrived in Cambridge in 1932 to carry out two year's postgraduate work. Although she says that she was too busy for much political activity, there is no doubt that Bernal's influence on her at that time and later was profound. As she stated: 'I myself came in [to science] during the 1930s and my guide book was Bernal's *Social Function of Science*.'[34] Bernal himself never practised 'pure' science. Since his childhood in Ireland he had been fascinated by science and saw it as the path to the liberation of his native country. For him, science always had a social purpose, about which he wrote in great profusion. Bernal entered the scientific world in the early twenties, beginning his scientific career as a mathematician and physicist, but became involved in crystallography in

PLATE 8 Portrait of Dorothy Hodgkin. A photograph taken in 1964 at the time of her being awarded the Nobel Prize for chemistry.

1923 at the Davy Faraday Laboratory. In this same year he became a Marxist, believing that it would provide the only framework in which science could become the key to human progress and survival. Bernal moved to the Cavendish Institute, Cambridge, in 1928, and published his first book, *The World, the Flesh and the Devil* in the following year. This was Bernal's vision of the future and was essentially a demand for a planned society in which the scientific method could be introduced into all areas of human activity. He believed that there would follow a great increase in man's understanding of himself and the universe, and freedom from war, famine and disease. However, although Bernal and other socialist scientists [like Needham and Haldane] equated the causes of science and socialism, in the twenties their scientific careers came well ahead of political commitments. It was only during the early thirties that they began to widen the scope of their political activities, a move prompted by both the general political situation and one particular event in 1931 – the second International Congress of the History of Science and Technology. The general situation in Britain following the economic crisis in 1929 was bleak.[35] For scientists the situation was impoverished; funding and research positions were scarce. Unemployment in general was high and poverty rife – half the nation was estimated as living below the nutritional level necessary for full health.[36]

There was a growing awareness amongst left-wing scientists of a need for the politicization of science. As Needham reflected: 'I tried to keep to my own field but politics would keep breaking in.'[37] The event which focused many of their ideas occurred in the summer of 1931. The second International Congress in the History of Science and Technology took place at the Science Museum, London and was unexpectedly attended by a delegation of eight Soviet politicians, administrators, scientists and historians. Headed by Nikolai Bukharin, the delegation also included the biologist Vavilov and the historian and physicist Boris Hessen. Although the papers presented by the Soviets were viewed by many of the historians present as sacrilegious, to scientists of the left, they were an inspiration which crystallized many of the problems they had encountered while trying to envision a scientific programme allied to society. Hessen's paper on Newton depicted the *Principia* as rooted in seventeenth-century social and economic life; it was represented as being the intellectual base of industrial capitalism. To an audience accustomed to perceiving science as a body of ideas fathered by scientific genius this concept appeared either laughable or sacrilegious. Bukharin himself argued that scientific theorists should be closely linked to those who apply science, and that because in capitalist societies the 'mental' and the 'manual' are totally divided this could never happen. According to Bukharin, in the USSR, 'We are arriving not only at a synthesis of science, but at a social synthesis of science and practice.'[38] We can see why the effect of this conference was so great for socialist scientists like Bernal.* Here was a historian and scientist depicting science as a product of society, not as a 'pure' discipline governed by its own internal logical development. This kind of analysis promised space and possibilities: if science is a product of a particular society, then by changing the society it should be possible to change the science and vice versa. In a socialist society science would be planned with specific aims in mind; for both economic progress and alleviation of human suffering; it would indeed become a 'science for the people'.

The Connections: Science, Politics and Mothering

The 1931 conference, then, provided the impetus for the creation of a 'popular front' of scientists, who took a radical stance with respect to the planning and aims of the scientific enterprise. Dorothy Hodgkin arrived in Cambridge in 1932 and became part of the social group led by Bernal. Already a socialist and pacifist, her beliefs were strengthened by the network in which she found herself. She attended meetings of the Association of Scientific Workers which had been recently resurrected by Bernal. She was also a regular attender of the Cambridge Scientists' Anti-War Group [CSAWG], which was formed by Bernal in 1932. When she returned to Oxford in 1934 to take up a research fellowship at Somerville she again became involved in political activity. She attended the Thursday Lunch Club meetings initiated by the socialist historian G. D. H. Cole. Visits to Cambridge included CSAWG meetings, which were also attended by other left-wing women scientists; including Dorothy Needham, Reinet Fremlin, Marjorie Stephenson and Nora Wooster. In 1937 Dorothy married Thomas Hodgkin, a socialist, who was an authority on African affairs and a lecturer in the Workers' Education Association. Their three children were born in the late thirties and early forties, and she was able to continue both political and scientific work with the help of an extended family at home.

That the political and scientific were and are inseparable for Dorothy is obvious throughout her life. Like Bernal, she was concerned with the application of science for the good of humanity and not for warfare. She was concerned with the planning of science, campaigning for more government support for research and also for encouraging women into science. However, her ideals and practice with respect to science differ from those of Bernal and other socialist scientists in specific ways which I suggest are gender based. The first of these is the stress she lays on co-operative effort in science, the second is involvement with the personal as well as the theory, and the last point concerns her pacifism.

For Dorothy, science was and still is a co-operative venture. She believes that scientific knowledge should be shared, not secretive, and should never be used for competition, profit or warfare. Her attitude is that it is the development of ideas and knowledge that is important rather than who develops them. 'In recent years the pace of discovery and application has enormously increased, the co-operative effects of very many hands.'[39] With respect to science at large Dorothy is idealistic in the claim that all scientific discoveries are made in a spirit of co-operation, but the concept is accurate when applied to her own work. The environment she created in her own laboratory was friendly and open to students and colleagues alike. Despite being a gloomy place, the laboratory was happy and homely, and Dorothy was always ready to share ideas and knowledge.

The claim that Dorothy is involved with the personal, not only with the theory, is substantiated throughout her writing. Even in scientific papers she alludes to her humanitarian reasons for following specific lines of research. For her, scientific research is about people, not just molecules; it is about the great events in medical history and the way those events have affected the sufferers. She wrote of diabetes and of the isolation of insulin by Banting and Best in 1921: 'Medical history is to the observer full of events that produce sudden and spectacular changes, before, many die, and afterwards and suddenly they live.'[40] Her work has been about people dying or not dying. Her crystallographic research centred on three molecules – insulin, vitamin B12 and penicillin.[41] There were possibilities for working on many crystals; according to Riley the Oxford laboratory was 'open house', a kind of odd-job workshop.[42] But the molecules Dorothy concentrated on were those of specific importance to medical science. They had a known therapeutic role in specific diseases – diabetes, infections and pernicious anaemia, all of which were fairly common and often fatal. Although the substances were already in clinical use, by elucidating the three-dimensional structure of the molecule by X-ray crystallography it would become possible to work out the mechanisms of action of normal and disease processes, and thus gain more insight into both therapy and prevention. As Dorothy said of insulin:

> It is effective, it keeps patients alive who would otherwise die. But its use is at best very painful, imperfect medicine. If we knew in all fundamental detail how insulin acts to control our metabolism we might be able to devise far better methods for treating the different disorders associated with diabetes, blindness for example, which insulin injections only very imperfectly control.[43]

She was thus concerned with patients' every-day lives and how her work would help improve their well-being. The influence of her own family life on her work is best shown by her attitude towards penicillin research in the early 1940s. According to Dorothy, she became involved in this research in 1942 at the request of Howard Florey,[44] and also because her insulin studies were proving very complicated owing to the molecule's size.[45] However, later she was to say of penicillin and other subsequently-found antibiotics:

> But I like best to see the effects in the general improvement of life expectation; for example the yearly average of deaths in early childhood from infectious diseases in 1931–5 was 3000 in this country; in 1966 it was 500. No doubt other factors than antibiotics contributed to this effect, but still the change is staggering.[46]

Dorothy's children were born and spent their infancy in the era before antibiotics, all were under five when she began work on penicillin. I suggest that the reason for her decision to work on penicillin was directly dependent on her involvement with her own children. Similarly, I suggest that her practice of pacifism and its integration into her scientific work was dependent on hopes and fears for her children's future.

It has been pointed out by Wersky,[47] that taking into account the overall ratio of women to men researchers in Cambridge during the 1930s, there were many more women in the CSAWG. He states that this cannot be explained away by 'a passing reference to women's traditional pacifism.' Traditional pacifism is rather a loose, emotive term; rather we should be basing the pacifism in social experience. As has been explained in the previous section, women's pacifism at this time was based on an essentialist ideal. Women, because of their nurturing and caring qualities were depicted as being in a privileged position with respect to pacifism.

They were the group who should be deeply and personally involved in making the world a safer place for their children. Peace groups at this time were not isolated little bands; in Cambridge there were many, including a local branch of the WILPF; and especially towards the end of the thirties these met for joint activities. For example, in December 1938, a peace week was held, including many meetings, play performances and a torchlight procession concluding on Parker's Piece. Therefore to find comparatively more women in CSAWG should not be surprising at all.

So how does this affect scientists' work? It is clear that Dorothy Hodgkin's work has exemplified pacifist ideals. She has warned of the need for vigilance in scientific practice to ensure its benevolent use. For example, she has said with respect to genetic engineering of proteins: 'Vigilance and caution are certainly necessary if these experiments are to do the good we hope they will and no harm. We have seen too many dangerous and undesirable effects derived from potentially very useful scientific observations to be complacent.'[48] Most importantly she has made certain that her own research has been carried out for the benefit of people. At the beginning of the Second World War, Dorothy was faced, along with many others, by a conflict between socialism and pacifism. She chose the middle ground, working on penicillin structure. Although this work was of great military significance, its immediate use in a world-wide influenza epidemic was of great importance, and it has since become what Dorothy would have referred to as 'one of the great medical advances of the twentieth century.'

Dorothy once said of Bernal that holding socialist beliefs implied that a scientist should work towards a perfect community and live according to socialist principles within the present, imperfect one.[49] As an interpretation of Dorothy's own life this is faultless. In her peace work she has worked towards a perfect community, and within the present, imperfect one she has endeavoured to improve life expectancy and the quality of that life.

DISCUSSION

To end I should like to discuss contemporary issues arising not from the woman question in science, but from what Sandra Harding has referred to as 'the science question in feminism'. If we think of the scientific enterprise as an academic discipline which is devoid of cultural significance or genesis, then there are only two dialogues open to women. The first is that although women have problems at present gaining equal opportunities and promotion, all that is required is that conditions should change so that women can take their rightful place in science. The second is that as science has been constructed by men and is aggressive and destructive, women should avoid the discipline completely. Both these dialogues are destructive; to partake in science results in imbibing its masculinity; to eschew science means ignoring and casting aside an enterprise which is central to our culture. The only way, therefore, to understand the science question in feminism is to perceive science for what it is – a cultural activity, which has been constructed by particular individuals and which, because it is a cultural activity, is therefore amenable to change.

As I have shown, it was possible for Dorothy Hodgkin to carry out her science in a way that not only affirmed her credibility as an acclaimed scientist, but also kept faith with her beliefs as pacifist, socialist and mother. However, a word of caution is needed here: scientists need networks and allies within the scientific profession if they are to change the way science is practised. Dorothy was part of a social group who believed that science was allied to social and political constructs, and also that a different kind of science could provide the answers to humanity's ills. While, for some, the aims of science are still perceived as the alleviation of human problems, the enterprise is now inextricably woven into an ethos of profit-making and nationalistic competition. Government education policies (for example, the national curriculum) will soon ensure that from primary school upwards students will imbibe the ideology of science for profit rather than science for the good of humanity. In a political climate which is becoming increasingly regimented towards money rather than people it is also becoming more difficult to stand up and be counted on the side of humanity and to criticize how and why science projects are undertaken. It is clear that not only do we as individuals need to question government control and indoctrination with respect to scientific enterprise, but that we also need to build up networks of awareness within the profession itself, which would enable working women scientists to criticize present policies and to work for policies which in turn will work for humanity.

This is especially important with regard to military science. Feminists and pacifists have tended to think of the military as devoid of social constructs, and since, for them, it is the fount of all evil, no further comment or analysis is required. However, since military values are part of our culture and tradition, it would appear that this is a naïve perspective, which in the end is self-defeating. Whilst the institution is thought of as devoid of social significance, although it is destructive and evil, it is also unassailable and sacrosanct. In the same way that perceiving science as outside society prevents us changing it, so perceiving war and the military as outside society prevents us from eradicating it.

Militaristic aims are at present portrayed as deterrence not attack, as protection not aggression; upholding an established military myth. If we take note of feminist analysis of the analogy between the public and domestic spheres, then this myth of protection raises certain questions with respect to women. Are we protected by men because we are their property? Are these the same men who commit atrocities against women and children simply because they do not belong to them? Are those men who portray themselves as our protectors the same men who become the perpetrators of violence against us at home? Following on from this, can we suggest that in a patriarchal society, where women are exploited in both the economic and reproductive spheres, their experience of alienation and violence become their 'natural state'; that in practice and theory military and domestic violence are two sides of the same coin? This suggestion, which appeared so obvious to me, has met with such opposition from male friends and colleagues that it would appear there is a necessity to try and understand the reasons why this was so. Could it be that while military violence is publicly sanctioned or 'justified', domestic violence, though illegitimate, is much more prevalent, and thus has gained a perverse form of social sanction? At the same time, domestic violence is publicly defined alongside rape, as deviant, so that ordinary men can perceive themselves as 'other'. If it is suggested that the roots of both military 'legitimate' and domestic 'illegitimate' violence are firmly situated in our society this presents male violence in a light which men may not care to acknowledge. It is these issues we need to address, to try and understand what protection and deterrence means in a contemporary context.

We need to see that as long as military aims are portrayed as deterrence, then militaristic science can be construed as the epitome of anti-war strategy and as a way for scientists to protect their own. Scientists thus have few qualms about working in these areas and can easily justify themselves doing so.

But pacifism, as historical treatment reveals, and as Jane Addams has observed, is not a negative ethos; not merely an anti-war stance. It is positively working for peace; thus a pacifist science would be a science which has its aims, projects and uses firmly rooted in the thirties slogan 'a science for the people'. It has to be an enterprise in which scientists engage knowingly; in other words, they have to recognize the enterprise for what it is, rather than by the myths that surround it. Finally, I have suggested that it is possible for women to bring a more pacifist and humane practice to science. Indeed, in the early decades of this century, it was women who met the pacifist challenge, organizing internationally in their attempts to bring peace to the world. Yet the burden for change need not in the future rest solely on women, and if we move away from an essentialist reading of women's nature there is no reason to preclude anyone from working towards peace.

NOTES

1 For example, in pre-war Germany, the Rector of Frankfurt University, Dr Ernst Kriek wrote in 1937: 'It is not objective science which is the purpose of our university training, but the heroic science of the soldier, the militant and fighting science, '*L'École Hitlerienne et l'etranger* (1937); quoted in J. D. Bernal, *The Social Function of Science* (Routledge, London, 1939), p. 218.
2 'Gender' used advisedly.
3 E.g. the Society for Social Responsibility in Science.
4 *News Bulletin*, Nov. 1930.
5 Brian Easlea, *Fathering the Unthinkable: Masculinity, Scientists and the Nuclear Arms Race* (Pluto Press, London, 1983).
6 M. Foucault, *Archaeology of Knowledge* (Tavistock, London, 1972), p. 17.
7 There is no doubt that many women scientists have done this; they have made a choice between family and career, and in choosing the latter have adopted 'male' values with respect to work and personal lives.
8 See Lynda Birke, 'Life' as We Have Known It', ch. 8, this vol.
9 Hilary Rose, 'Hand, brain and heart. A feminist epistemology for the natural sciences' in S. Harding and J. O'Barr (eds), *Sex and Scientific Enquiry* (University of Chicago Press, 1987).
10 H. M. Collins, *Changing Order; Replication and Induction in Scientific Practice* (Sage, London, 1985).
11 R. Lewontin, '"Honest Jim" Watson's big thriller about D.N.A.' (1968) in James D. Watson, *The Double Helix*, ed. Gunter S. Stent (Weidenfeld and Nicholson, London, 1981), p. 186.
12 J. Addams, quoted in Gertrude Bussey and Margaret Tims, *W.I.L.P.F. 1915–1965* (Allen and Unwin, London, 1965), frontispiece.

13 J. S. Hallowes, *Mothers of Men and Militarism* (Headley Brothers, London, n.d.), p. 18.

14 A third meeting was due to be held in 1915, but was abandoned.

15 The NUWSS was the non-militant suffrage organization led by Millicent Garrett Fawcett. Members, referred to as suffragists (also nicknamed 'Polites' by their more militant sisters), included Helena Swanwick, Margaret Ashton, Catherine Marshall, Maude Roydon, Emmeline Pethick-Lawrence, Crystal Macmillan and Kathleen Courtney. During the First World War, the division between the NUWSS and the militant suffragettes (the Women's Social and Political Union, WSPU) deepened and fragmented the suffrage movement. WSPU was led by Emmeline and Christabel Pankhurst, who supported the war effort. Even some of the non-militant NUWSS members eventually came to support the war.

16 Rosika Schwimmer was a radical feminist who had been president of the National Association of Women Office Workers in Budapest and was well known in the international suffrage movement.

17 Some German women and all French and Russian women were prevented by their governments from attending the conference. Of the original British contingent of 180, eventually only 3 women (Crystal Macmillan, Kathleen Courtney and Emmeline Pethick-Lawrence) managed to attend.

18 Member organizations of the Peace and Disarmament Committee of the Women's International Organizations in 1937 included 16 separate groups.

19 Else Zenthen (WILPF International Chairwomen) in Bussey and Tims, *W.I.L.P.F. 1915–1965*, frontispiece.

20 Jane Addams, an American suffragist, was the first president of WILPF: quoted in ibid.

21 Madame Clara Guthrie D'Arcis [president of the Union de la Femme pour la Concorde Internationale] in 'The prevention of the causes of war – addresses delivered at the International Council of Women's Conference' at Wembley, May 1924, (International Council of Women, London, 1924), p. 12.

22 Maude Roydon had joined the NUWSS in 1908 and edited the Union's journal, *The Common Cause*. She was a devout Christian and became assistant preacher at the City Temple, London, in 1917.

23 G. P. Gooch in *The Teaching of History and Scripture*, report of a conference arranged by the Women's International Committee for Permanent Peace, Westminster, Jan. 1917, p. 10.

24 Agnes MacPhail, *Pax International*, Aug. 1927.

25 *W.I.L.P.F. Monthly News Sheet*, Jan./Feb. 1928.

26 For example, Mrs Snowden, wife of Phillip Snowden, who was chancellor of the Exchequer in 1930; and Emmeline Pethick-Lawrence, whose husband was Under-Secretary of Finance in the Labour government of 1929.

27 *Pax International*, Feb. 1926.

28 Marcell Capy, *Pax International*, July 1929.

29 For a contemporary interpretation of maternal and pacifist practice see Sarah Ruddick, 'Preservative love and military destruction' in J. Trebilcot (ed.), *Mothering, Essays in Feminist Theory* (Rowman and Allenheld, New Jersey, 1984).

30 'The prevention of the causes of war', International Council of Women's Conference, Wembley, 1924 (International Council of Women, London, 1924), p. 32.

31 Max Perutz, 'Forty Years of Friendship with Dorothy' in G. Dodson, J. Glusker and D. Sayre (eds), *Structural Studies on Molecules of Biological Interest* (Oxford, Clarendon Press, 1981), p. 6.

32 Caroline Moorehead, 'Profile of Dorothy Hodgkin', *The Times*, 21 April 1975.

33 Personal interview with Dorothy Hodgkin at Park Parade, Cambridge, in 1986.

34 D. Hodgkin, 'Discoveries and their uses', Presidential Address, (British Association for the Advancement of Science, 1978), p. 1.

35 For left-wing supporters, the situation deteriorated in 1931 when the Labour MP Ramsay MacDonald formed a coalition government, resulting in many socialists becoming disillusioned with the official party.

36 This point was especially important to many biochemists working on nutritional research.

37 J. Needham, *Time: The Refreshing River* (London, 1943), p. 11.

38 N. Bukharin et al, in *Science at the Crossroads* (London, 1971), p. 31.

39 Hodgkin, 'Discoveries and their uses', p. 9.

40 D. Hodgkin, 'X-rays and the structure of insulin', *British Medical Journal*, 4 (1971), p. 447.

41 Dorothy received the Nobel Prize for work carried out on vitamin B12 in the 1950s.

42 D.P. Riley, 'Oxford, the early years', in Dodson, Glusker and Sayres (eds), *Structural Studies*, p. 25.

43 Hodgkin, 'Discoveries and their uses', p. 10.

44 See further J. C. Sheehan, *The Enchanted Ring, The Untold Story of Penicillin* (MIT Press, Cambridge, Mass., 1984).

45 The structure of penicillin was thought to be easier to solve as it was smaller and less complex.

46 Hodgkin, 'Discoveries and their uses', p. 7.

47 G. Wersky, *The Visible College* (Allen Lane, London, 1978).

48 Hodgkin, 'Discoveries and their uses', p. 19.

49 D. Hodgkin, 'Obituary of J. D. Bernal', *Biographical Memoirs of the Fellows of the Royal Society*, 26 (1980), pp. 17–84.

Additional Reading

Barker-Benfield, C. J., *The Culture of sensibility, sex and society in eighteenth century Britain*, (University of Chicago Press, Chicago, 1992).

Bequaert-Holmes, Helen and Purdy, Laura M., *Feminist Perspectives in Medical Ethics*, (Indiana University Press, Bloomington and Indianopolis, 1992).

Fox-Keller, Evelyn, *Secrets of Life, Secrets of Death, essays on language, gender and science*, (Routledge, London 1992).

Haraway, Donna, *Simians, Cyborgs, and Women: the reinvention of nature*, (Free Association Press, London, 1991).

Harding, Sandra, *Whose Science? Whose Knowledge?*, (Open University Press, Milton Keynes, 1991).

Kickup, Gill and Smith-Keller, Laurie, eds, *Inventing Women: science, technology and gender*, (Polity Press in association with Open University Press, Cambridge, 1992).

Laqueur, *Making Sex: Body & Gender from the Greeks to Freud*, (Harvard University Press, Cambridge, Mass., and London, 1990).

Rosser, Sue V., *Biology and Feminism: A Dynamic Interaction*, (Twayne Publishers, New York, 1992).

Schiebinger, Londa, *Nature's Body: Gender in the Making of Modern Science*, (Beacon Press, Boston, 1993).

Wajcman, Judy, *Feminism Confronts Technology*, (Polity Press: Cambridge, 1991).

Zuckerman, Harriet, Cole, Johnathan R. and Brueu, John T., eds, *The Outer Circle: Women in the Scientific Community*, (W.W. Norton & Company, New York and London, 1991).

Index